Transportation Conformity Guidance for Quantitative Hot-Spot Analyses in PM$_{2.5}$ and PM$_{10}$ Nonattainment and Maintenance Areas

Appendices A-K

Transportation and Climate Division
Office of Transportation and Air Quality
U.S. Environmental Protection Agency

United States
Environmental Protection
Agency

EPA-420-B-13-053
November 2013

Appendix A:
Clearinghouse of Websites, Guidance, and Other Technical Resources for PM Hot-spot Analyses

A.1 INTRODUCTION

This appendix is a centralized compilation of documents and websites referenced in the guidance, along with additional technical resources that may be of use when completing quantitative PM hot-spot analyses. Refer to the appropriate sections of the guidance for complete discussions on how to use these resources in the context of completing a quantitative PM hot-spot analysis. The references listed are current as of this writing; readers are reminded the check for the latest versions when using them for a particular PM hot-spot analysis.

A.2 TRANSPORTATION CONFORMITY AND CONTROL MEASURE GUIDANCE

The EPA hosts an extensive library of transportation conformity guidance online at: www.epa.gov/otaq/stateresources/transconf/policy.htm (unless otherwise noted). The following specific guidance documents, in particular, may be useful references when implementing PM hot-spot analyses:

- "Policy Guidance on the Use of MOVES2010 for SIP Development and Transportation Conformity, and Other Purposes," EPA-420-B-09-046 (December 2009). This document describes how and when to use the MOVES2010 emissions model for SIP development, transportation conformity determinations, and other purposes.

- "EPA Releases MOVES2010 Mobile Source Emissions Model: Questions and Answers," EPA-420-F-09-073 (December 2009).

- "EPA Releases MOVES2010a Mobile Source Emissions Model Update: Questions and Answers," EPA-420-F-10-050 (August 2010).

- "Technical Guidance on the Use of MOVES2010 for Emission Inventory Preparation in State Implementation Plans and Transportation Conformity," EPA-420-B-10-023 (December 2009). This document provides guidance on appropriate input assumptions and sources of data for the use of MOVES2010 in SIP submissions and regional emissions analyses for transportation conformity purposes.

- EPA and FHWA, "Transportation Conformity Guidance for Qualitative Hot-spot Analyses in $PM_{2.5}$ and PM_{10} Nonattainment and Maintenance Areas," EPA-420-B-06-902 (March 2006).

- EPA and FHWA, "Guidance for the Use of Latest Planning Assumptions in Transportation Conformity Determinations," EPA-420-B-08-901 (December 2008).

- "Guidance for Developing Transportation Conformity State Implementation Plans," EPA-420-B-09-001 (January 2009).

- EPA-verified anti-idle technologies (including technologies that pertain to trucks) can be found at: www.epa.gov/otaq/smartway/transport/what-smartway/verified-technologies.htm#idle.

- For additional information about quantifying the benefits of retrofitting and replacing diesel vehicles and engines for conformity determinations, see EPA's website for the most recent guidance on this topic: www.epa.gov/otaq/stateresources/transconf/policy.htm.

- For additional information about quantifying and using long duration truck idling benefits for conformity determinations, see EPA's website for the most recent guidance on this topic: www.epa.gov/otaq/stateresources/transconf/policy.htm.

FHWA's transportation conformity site has additional conformity information, including examples of qualitative PM hot-spot analyses. Available at: www.fhwa.dot.gov/environment/air_quality/conformity/practices/.

A.3 MOVES MODEL TECHNICAL INFORMATION AND USER GUIDES

MOVES, any future versions of the model, the latest user guides, and technical information can be found at www.epa.gov/otaq/models/moves/index.htm, including the following:

- "User Guide for MOVES2010a." This guide provides detailed instructions for setting up and running MOVES2010a. Available at www.epa.gov/otaq/models/moves/index.htm.

Policy documents and Federal Register announcements related to the MOVES model can be found on the EPA's website at: www.epa.gov/otaq/stateresources/transconf/policy.htm#models.

Guidance on using the MOVES model at the project level, as well as illustrative examples of using MOVES for quantitative PM hot-spot analyses, can be found in Section 4 of the guidance and in Appendices D, E and F.

A.4 EMFAC2007 MODEL TECHNICAL INFORMATION, USER GUIDES, AND OTHER GUIDANCE

EMFAC2007, its user guides, and any future versions of the model can be downloaded from the California Air Resources Board website at: www.arb.ca.gov/msei/onroad/latest_version.htm.

Policy documents and Federal Register announcements related to the EMFAC model can be found on the EPA's website at: www.epa.gov/otaq/stateresources/transconf/policy.htm#models.

Supporting documentation for EMFAC, including the technical memorandum "Revision of Heavy Heavy-Duty Diesel Truck Emission Factors and Speed Correction Factors" cited in Section 5 of this guidance, can be found at www.arb.ca.gov/msei/supportdocs.htm#onroad.

Instructions on using the EMFAC model at the project level, as well as examples of using EMFAC for quantitative PM hot-spot analyses, can be found in Section 5 of the guidance and in Appendices G and H.

A.5 DUST EMISSIONS METHODS AND GUIDANCE

Information on calculating emissions from paved roads, unpaved roads, and construction activities can be found in AP-42, Chapter 13 (Miscellaneous Sources). AP-42 is EPA's compilation of data and methods for estimating average emission rates from a variety of activities and sources from various sectors. Refer to EPA's website to access the latest versions of AP-42 sections and for more information about AP-42 in general: www.epa.gov/ttn/chief/ap42/index.html.

Guidance on calculating dust emissions for PM hot-spot analyses can be found in Section 6 of the guidance.

A.6 LOCOMOTIVE EMISSIONS GUIDANCE

The following guidance documents, unless otherwise noted, can be found on or through the EPA's locomotive emissions website at: www.epa.gov/otaq/locomotives.htm:

- "Procedure for Emission Inventory Preparation - Volume IV: Mobile Sources," Chapter 6. Available online at: www.epa.gov/OMS/invntory/r92009.pdf. Note that the emissions factors listed in Volume IV have been superseded by the April 2009 publication listed below for locomotives certified to meet EPA standards.

- "Emission Factors for Locomotives," EPA-420-F-09-025 (April 2009). Available online at: www.epa.gov/otaq/regs/nonroad/locomotv/420f08014.htm.

- "Control of Emissions from Idling Locomotives," EPA-420-F-08-014 (March 2008).

- "Guidance for Quantifying and Using Long Duration Switch Yard Locomotive Idling Emission Reductions in State Implementation Plans," EPA-420-B-04-002 (January 2004). Available online at: www.epa.gov/otaq/smartway/documents/420b04002.pdf.

- EPA-verified anti-idle technologies (including technologies that pertain to locomotives) can be found at: www.epa.gov/otaq/smartway/transport/what-smartway/verified-technologies.htm#idle.

Guidance on calculating locomotive emissions for PM hot-spot analyses can be found in Section 6 of the guidance and in Appendix I.

A.7 AIR QUALITY DISPERSION MODEL TECHNICAL INFORMATION AND USER GUIDES

The latest version of "Guideline on Air Quality Models" (Appendix W to 40 CFR Part 51) (dated 2005 as of this writing) can be found on EPA's SCRAM website at: www.epa.gov/scram001/guidance_permit.htm.

Both AERMOD and CAL3QHCR models and related documentation can be obtained through EPA's Support Center for Regulatory Air Models (SCRAM) web site at: www.epa.gov/scram001. In particular, the following guidance may be useful when running these models:

- AERMOD Implementation Guide

- AERMOD User Guide ("User's Guide for the AMS/EPA Regulatory Model – AERMOD")

- CAL3QHCR User Guide ("User's Guide to CAL3QHC Version 2.0: A Modeling Methodology for Predicting Pollutant Concentrations Near Roadway Intersections")

- MPRM User Guide

- AERMET User Guide

Information on locating and considering air quality monitoring sites can be found in 40 CFR Part 58 (Ambient Air Quality Surveillance), particularly in Appendices D and E to that part.

Guidance on selecting and using an air quality model for quantitative PM hot-spot analyses can be found in Sections 7 and 8 of the guidance and in Appendix J. Illustrative examples of using an air quality model for a PM hot-spot analysis can be found in Appendices E and F.

A.8 TRANSPORTATION DATA AND MODELING CONSIDERATIONS

The following is a number of technical resources on transportation data and modeling which may help implementers determine the quality of their inputs and the sensitivity of various data.

A.8.1 Transportation model improvement

The FHWA Travel Model Improvement Program (TMIP) provides a wide range of services and tools to help planning agencies improve their travel analysis techniques. Available online at: http://tmip.fhwa.dot.gov/.

A.8.2 Speed

"Evaluating Speed Differences between Passenger Vehicles and Heavy Trucks for Transportation-Related Emissions Modeling." Available online at: www.ctre.iastate.edu/reports/truck_speed.pdf.

A.8.3 Project level planning

"NCHRP 255: Highway Traffic Data for Urbanized Area Project Planning and Design." Available online at: http://tmip.fhwa.dot.gov/sites/tmip.fhwa.dot.gov/files/NCHRP_255.pdf.

A.8.4 Traffic analysis

Traffic Analysis Toolbox website: http://ops.fhwa.dot.gov/trafficanalysistools/.

"Traffic Analysis Toolbox Volume I: Traffic Analysis Tools Primer." Federal Highway Administration, FHWA-HRT-04-038 (June 2004). Available online at: http://ops.fhwa.dot.gov/trafficanalysistools/tat_vol1/vol1_primer.pdf.

The Highway Capacity Manual Application Guidebook. Transportation Research Board, Washington, D.C., 2003. Available online at: http://hcmguide.com/.

The Highway Capacity Manual 2000. Transportation Research Board, Washington, D.C., 2000. Not available online; purchase information available at: http://144.171.11.107/Main/Public/Blurbs/Highway_Capacity_Manual_2000_152169.aspx. As of this writing, the 2000 edition is most current; the most recent version of the manual, and the associated guidebook, should be consulted when completing PM hot-spot analyses.

Appendix B:
Examples of Projects of Local Air Quality Concern

B.1 INTRODUCTION

This appendix gives additional guidance on what types of projects may be projects of local air quality concern requiring a quantitative PM hot-spot analysis under 40 CFR 93.123(b)(1). However, as noted elsewhere in this guidance, PM_{10} nonattainment and maintenance areas with approved conformity SIPs that include PM_{10} hot-spot provisions from previous rulemakings must continue to follow those approved conformity SIP provisions until the SIP is revised; see Appendix C for more information.

B.2 EXAMPLES OF PROJECTS THAT REQUIRE PM HOT-SPOT ANALYSES

EPA noted in the March 2006 final rule that the examples below are considered to be the most likely projects that would be covered by 40 CFR 93.123(b)(1) and require a $PM_{2.5}$ or PM_{10} hot-spot analysis (71 FR 12491).[1]

Some examples of projects of local air quality concern that would be covered by 40 CFR 93.123(b)(1)(i) and (ii) are:
- A project on a new highway or expressway that serves a significant volume of diesel truck traffic, such as facilities with greater than 125,000 annual average daily traffic (AADT) and 8% or more of such AADT is diesel truck traffic;
- New exit ramps and other highway facility improvements to connect a highway or expressway to a major freight, bus, or intermodal terminal;
- Expansion of an existing highway or other facility that affects a congested intersection (operated at Level-of-Service D, E, or F) that has a significant increase in the number of diesel trucks; and,
- Similar highway projects that involve a significant increase in the number of diesel transit busses and/or diesel trucks.

Some examples of projects of local air quality concern that would be covered by 40 CFR 93.123(b)(1)(iii) and (iv) are:
- A major new bus or intermodal terminal that is considered to be a "regionally significant project" under 40 CFR 93.101[2]; and,

---

[1] EPA also clarified 93.123(b)(1)(i) in the January 24, 2008 final rule (73 FR 4435-4436).

[2] 40 CFR 93.101 defines a "regionally significant project" as "a transportation project (other than an exempt project) that is on a facility which serves regional transportation needs (such as access to and from the area outside of the region, major activity centers in the region, major planned developments such as new retail malls, sports complexes, etc., or transportation terminals as well as most terminals themselves) and would normally be included in the modeling of a metropolitan area's transportation network, including at a minimum all principal arterial highways and all fixed guideway transit facilities that offer an alternative to regional highway travel."

- An existing bus or intermodal terminal that has a large vehicle fleet where the number of diesel buses increases by 50% or more, as measured by bus arrivals.

A project of local air quality concern covered under 40 CFR 93.123(b)(1)(v) could be any of the above listed project examples.

B.3 EXAMPLES OF PROJECTS THAT DO NOT REQUIRE PM HOT-SPOT ANALYSES

The March 2006 final rule also provided examples of projects that would not be covered by 40 CFR 93.123(b)(1) and would not require a $PM_{2.5}$ or PM_{10} hot-spot analysis (71 FR 12491).

The following are examples of projects that are not a local air quality concern under 40 CFR 93.123(b)(1)(i) and (ii):
- Any new or expanded highway project that primarily services gasoline vehicle traffic (i.e., does not involve a significant number or increase in the number of diesel vehicles), including such projects involving congested intersections operating at Level-of-Service D, E, or F;
- An intersection channelization project or interchange configuration project that involves either turn lanes or slots, or lanes or movements that are physically separated. These kinds of projects improve freeway operations by smoothing traffic flow and vehicle speeds by improving weave and merge operations, which would not be expected to create or worsen PM NAAQS violations; and,
- Intersection channelization projects, traffic circles or roundabouts, intersection signalization projects at individual intersections, and interchange reconfiguration projects that are designed to improve traffic flow and vehicle speeds, and do not involve any increases in idling. Thus, they would be expected to have a neutral or positive influence on PM emissions.

Examples of projects that are not a local air quality concern under 40 CFR 93.123(b)(1)(iii) and (iv) would be:
- A new or expanded bus terminal that is serviced by non-diesel vehicles (e.g., compressed natural gas) or hybrid-electric vehicles; and,
- A 50% increase in daily arrivals at a small terminal (e.g., a facility with 10 buses in the peak hour).

Appendix C:
Hot-Spot Requirements for PM_{10} Areas with Pre-2006 Approved Conformity SIPs

C.1 INTRODUCTION

This appendix describes what projects require a quantitative PM_{10} hot-spot analysis in those limited cases where a state's approved conformity SIP is based on pre-2006 conformity requirements.[1] The March 10, 2006 final hot-spot rule defined the current federal conformity requirements for what projects require a PM hot-spot analysis (i.e., only certain highway and transit projects that involve significant levels of diesel vehicle traffic or any other project identified in the PM SIP as a local air quality concern).[2] However, there are some PM_{10} nonattainment and maintenance areas where PM_{10} hot-spot analyses are required for different types of projects, as described further below.

This appendix will be relevant for only a limited number of PM_{10} nonattainment and maintenance areas with pre-2006 approved conformity SIPs. This appendix is not relevant for any $PM_{2.5}$ nonattainment or maintenance areas, since the current federal $PM_{2.5}$ hot-spot requirements apply in all such areas. Project sponsors can use the interagency consultation process to verify applicable requirements before beginning a quantitative PM_{10} hot-spot analysis.

C.2 PM_{10} AREAS WHERE THE PRE-2006 HOT-SPOT REQUIREMENTS APPLY

Prior to the March 2006 final rule, the federal conformity rule required some type of hot-spot analysis for all non-exempt federally funded or approved projects in PM_{10} nonattainment and maintenance areas. These pre-2006 requirements are in effect for those states with an approved conformity SIP that includes the pre-2006 hot-spot requirements.

In PM_{10} areas with approved conformity SIPs that include the pre-2006 hot-spot requirements, a <u>quantitative</u> PM_{10} hot-spot analysis is required for the following types of projects:
- Projects which are located at sites at which PM_{10} NAAQS violations have been verified by monitoring;
- Projects which are located at sites which have vehicle and roadway emission and dispersion characteristics that are essentially identical to those of sites

---

[1] A "conformity SIP" includes a state's specific criteria and procedures for certain aspects of the transportation conformity process (40 CFR 51.390).
[2] See Section 2.2 and Appendix B of this guidance and the preamble of the March 2006 final rule (71 FR 12491-12493).

with verified violations (including sites near one at which a violation has been monitored); and

- New or expanded bus and rail terminals and transfer points which increase the number of diesel vehicles congregating at a single location.

This guidance should be used to complete any quantitative PM_{10} hot-spot analyses.

In addition, a <u>qualitative</u> PM_{10} hot-spot analysis is required in the pre-2006 hot-spot requirements for all other non-exempt federally funded or approved projects. For such analyses, consult the 2006 EPA-FHWA qualitative hot-spot guidance.[3]

These pre-2006 hot-spot requirements continue to apply in PM_{10} areas with approved conformity SIPs that include them until the state acts to change the conformity SIP. The conformity rule at 40 CFR 51.390 states that conformity requirements in approved conformity SIPs "remain enforceable until the state submits a revision to its [conformity SIP] to specifically remove them and that revision is approved by EPA."

C.3   REVISING A CONFORMITY SIP

EPA strongly encourages affected states to revise pre-2006 provisions and take advantage of the streamlining flexibilities provided by the current Clean Air Act. EPA's January 2008 final conformity rule significantly streamlined the requirements for conformity SIPs in 40 CFR 51.390.[4] As a result, conformity SIPs are now required to include only three provisions (consultation procedures and procedures regarding written commitments) rather than all of the provisions of the federal conformity rule.

EPA recommends that states with pre-2006 PM_{10} hot-spot requirements in their conformity SIPs act to revise them to reduce the number of projects where a hot-spot analysis is required. In affected PM_{10} areas, the current conformity rule's PM_{10} hot-spot requirements at 40 CFR 93.123(b)(1) and (2) will be effective only when a state either:

- Withdraws the existing provisions from its approved conformity SIP and EPA approves this SIP revision, or
- Revises its approved conformity SIP consistent with the requirements found at 40 CFR 93.123(b) and EPA approves this SIP revision.

---

[3] "Transportation Conformity Guidance for Qualitative Hot-spot Analyses in $PM_{2.5}$ and PM_{10} Nonattainment and Maintenance Areas," EPA420-B-06-902, found on EPA's website at: www.epa.gov/otaq/stateresources/transconf/policy/420b06902.pdf.
[4] "Transportation Conformity Rule Amendments to Implement Provisions Contained in the 2005 Safe, Accountable, Flexible, Efficient Transportation Equity Act: A Legacy for Users (SAFETEA-LU); Final Rule," 73 FR 4420.

Affected states should contact their EPA Regional Office to proceed with one of these two options. For more information about conformity SIPs, see EPA's "Guidance for Developing Transportation Conformity State Implementation Plans (SIPs)," EPA-420-B-09-001 (January 2009); available online at:
www.epa.gov/otaq/stateresources/transconf/policy/420b09001.pdf.

Appendix D:
Characterizing Intersection Projects for MOVES

D.1 INTRODUCTION

This appendix expands upon the discussion in Section 4.2 on how best to characterize links when modeling an intersection project using MOVES. The MOVES emissions model allows users to represent intersection traffic activity with a higher degree of sophistication compared to previous models. This appendix provides several options to describe vehicle activity to take advantage of the capabilities MOVES offers to complete more accurate PM hot-spot analyses of intersection projects. MOVES is the approved emissions model for PM hot-spot analyses in areas outside of California.

Exhibit D-1 is an example of a simple signalized intersection showing the links developed by a project sponsor to represent the two general categories of vehicle activity expected to take place at this intersection (approaching the intersection and departing the intersection).

Exhibit D-1. Example of Approach and Departure Links for a Simple Intersection

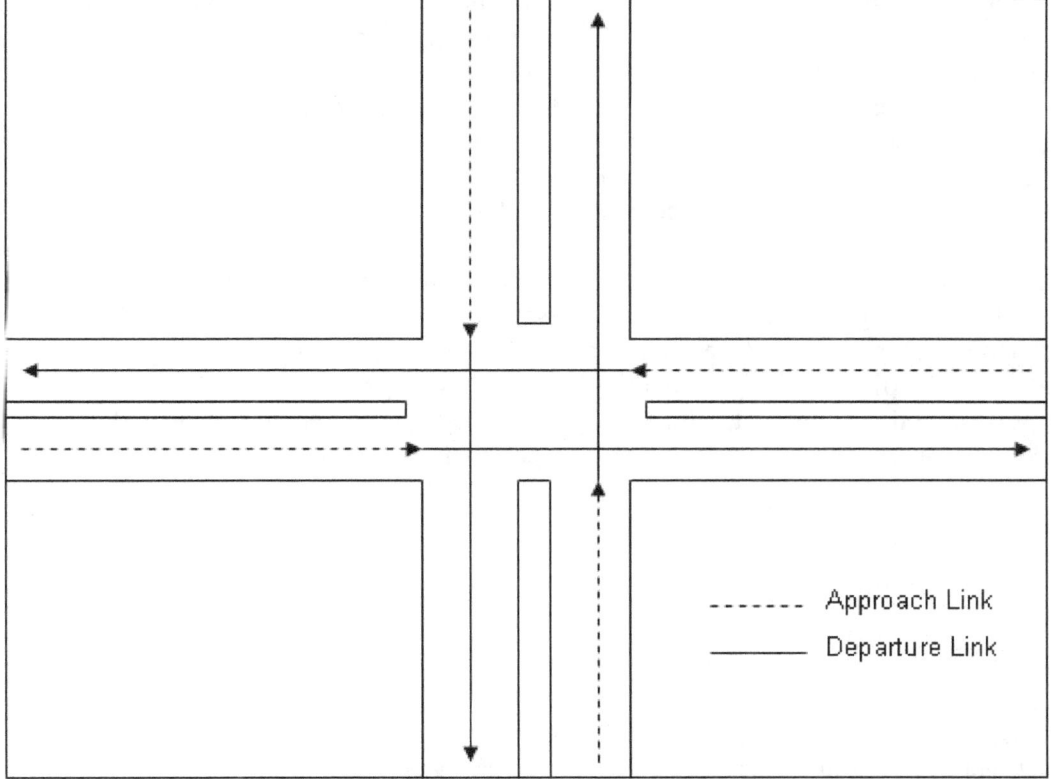

When modeling an intersection, each approach link or departure link can be modeled as one or more links in MOVES depending on the option chosen to enter traffic activity. This guidance suggests three possible options for characterizing activity on each approach and departure link (such as those shown in Exhibit D-1):

- Option 1: Using average speeds
- Option 2: Using link drive schedules
- Option 3: Using Op-Mode distributions

While Option 1 may need to be relied upon more during the initial transition to using MOVES, as more detailed data are available to describe vehicle activity, users are encouraged to consider using the Options 2 and 3 to take full advantage of the capabilities of MOVES.

Once a decision has been made on how to characterize links, users should continue to develop the remaining MOVES inputs as discussed in Section 4 of the guidance.

D.2 OPTION 1: USING AVERAGE SPEEDS

The first option is for the user to estimate the average speeds for each link in the intersection based on travel time and distance. Travel time should account for the total delay attributable to traffic signal operation, including the portion of travel when the light is green and the portion of travel when the light is red. The effect of a traffic signal cycle on travel time includes deceleration delay, move-up time in a queue, stopped delay, and acceleration delay. Using the intersection example given in Exhibit D-1, each approach link would be modeled as one link to reflect the higher emissions associated with vehicle idling through lower speeds affected by stopped delay; each departure link would be modeled as one link to reflect the higher emissions associated with vehicle acceleration through lower speeds affected by acceleration delay.

Project sponsors can determine congested speeds by using appropriate methods based on best practices for highway analyses. Some resources are available through FHWA's Travel Model Improvement Program (TMIP).[1] Methodologies for computing intersection control delay are provided in the Highway Capacity Manual.[2] All assumptions, methods, and data underlying the estimation of average speeds and delay should be documented as part of the PM hot-spot analysis.

---

[1] See FHWA's TMIP website: http://tmip.fhwa.dot.gov/.
[2] Users should consult the most recent version of the Highway Capacity Manual. As of the release of this guidance, the latest version is the *Highway Capacity Manual 2000*, which can be obtained from the Transportation Research Board (see http://144.171.11.107/Main/Public/Blurbs/152169.aspx for details).

D.3 OPTION 2: USING LINK DRIVE SCHEDULES

A more refined approach is to enter vehicle activity into MOVES as a series of link drive schedules to represent individual segments of cruise, deceleration, idle, and acceleration of a congested intersection. A link drive schedule defines a speed trajectory to represent the entire vehicle fleet via second-by-second changes in speed and highway grade. Unique link drive schedules can be defined to describe types of vehicle activity that have distinct emission rates, including cruise, deceleration, idle, and acceleration.

Exhibit D-2 illustrates why using this more refined approach can result in a more detailed emissions analysis. This exhibit shows the simple trajectory of a single vehicle approaching an intersection during the red signal phase of a traffic light cycle. This trajectory is characterized by several distinct phases (a steady cruise speed, decelerating to a stop for the red light, idling during the red signal phase, and accelerating when the light turns green). In contrast, the trajectory of a single vehicle approaching an intersection during the green signal phase of a traffic light cycle is characterized by a more or less steady cruise speed through the intersection.

Exhibit D-2. Example Single Vehicle Speed Trajectory Through a Signalized Intersection

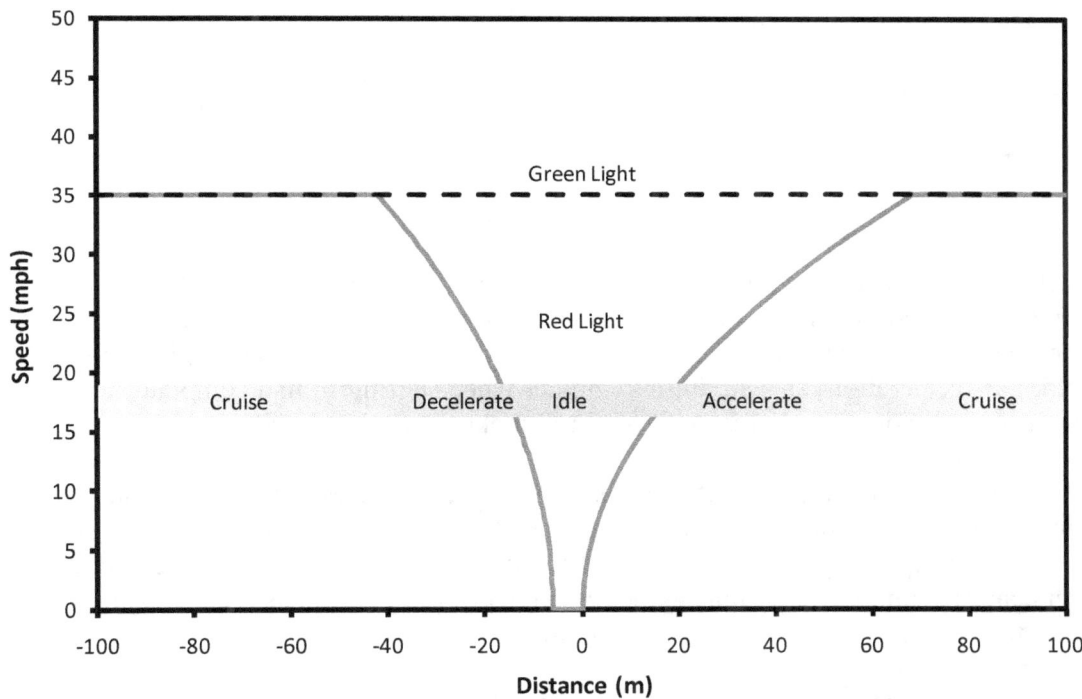

For the example intersection in Exhibit D-1, link drive schedules representing the different operating modes of vehicle activity on the approach and departure links can be determined. For approach links, the length of a vehicle queue is dependent on the number of vehicles subject to stopping at a red signal. Vehicles approaching a red traffic signal decelerate over a distance extending from the intersection stop line back to the stopping distance required for the last vehicle in the queue. The average stopping distance can be calculated from the average deceleration rate and the average cruise speed. Similarly, for the departure links, vehicles departing a queue when the light turns green accelerate over a distance extending from the end of the vehicle queue to the distance required for the first vehicle to reach the cruise speed, given the rate of acceleration and cruise speed. Exhibit D-3 provides an illustration of how the different vehicle operating modes may be apportioned spatially near this signalized intersection.

Exhibit D-3. Example Segments of Vehicle Activity Near a Signalized Intersection

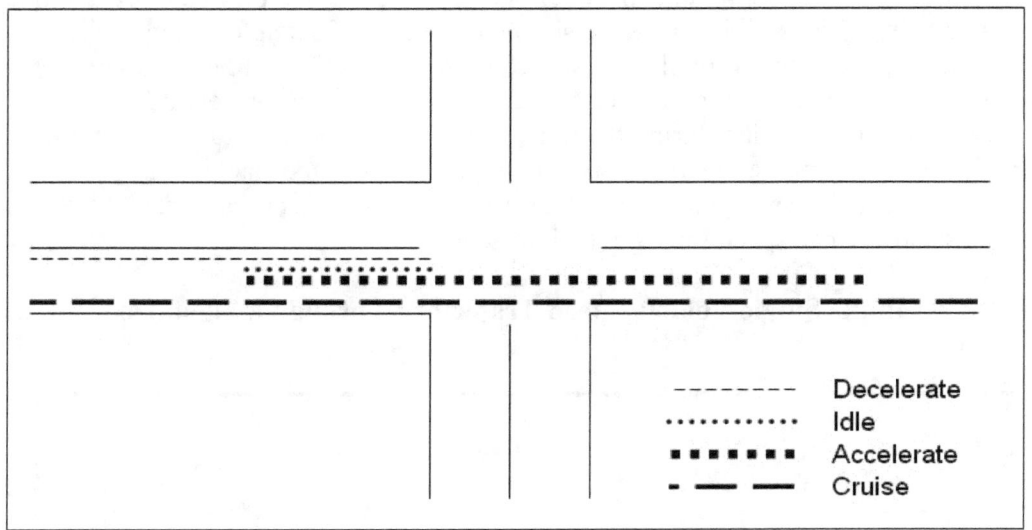

There are other considerations with numerous vehicles stopping and starting at an intersection over many signal cycles during an hour. For instance, heavy trucks decelerate and accelerate at slower rates than passenger cars. Drivers tend not to decelerate at a constant rate, but through a combination of coasting and light and heavy braking. Acceleration rates are initially higher when starting from a complete stop at an intersection, becoming progressively lower to make a smooth transition to cruise speed.

In the case of an uncongested intersection, the rates of vehicles approaching and departing the intersection are in equilibrium. Some vehicles may slow, and then speed up to join the dissipating queue without having to come to a full stop. Once the queue clears, approaching vehicles during the remainder of the green phase of the cycle will cruise through the intersection virtually unimpeded.

In the case of a congested intersection, the rate of vehicles approaching the intersection is greater than the rate of departure, with the result that no vehicle can travel through without stopping; vehicles approaching the traffic signal, whether it is red or green, will have to come to a full stop and idle for one or more cycles before departing the intersection. The latest Highway Capacity Manual is a good source of information for vehicle operation through signalized intersections. All assumptions, methods, and data underlying the development of link drive schedules should be documented as part of the PM hot-spot analysis.

The MOVES emission factors for each segment of vehicle activity obtained via individual link drive schedules are readily transferable to either AERMOD or CAL3QHCR, as discussed further in Section 7 of the guidance. There will most likely be a need to divide the cruise and the acceleration segments to account for differences in approach and departure traffic volumes.

Note: For both free-flow highway and intersection links, users may directly enter output from traffic simulation models in the form of second-by-second individual vehicle trajectories. These vehicle trajectories for each road segment can be input into MOVES using the Link Drive Schedule Importer and defined as unique LinkIDs. There are no limits in MOVES as to how many links can be defined; however, model run times increase as the user defines more links. A representative sampling of vehicles can be used to model higher volume segments by adjusting the resulting sum of emissions to account for the higher traffic volume. For example, if a sampling of 5,000 vehicles (5,000 links) was used to represent the driving patterns of 150,000 vehicles, then the sum of emissions would be adjusted by a factor of 30 to account for the higher traffic volume (i.e., 150,000 vehicles/5,000 vehicles). Since the vehicle trajectories include idling, acceleration, deceleration, and cruise, separate roadway links do not have to be explicitly defined to show changes in driving patterns. The sum of emissions from each vehicle trajectory (LinkID) represents the total emission contribution of a given road segment.

D.4 OPTION 3: USING OP-MODE DISTRIBUTIONS

A third option is for a user to generate representative Op-Mode distributions for approach and departure links by calculating the fraction of fleet travel times spent in each mode of operation. For any given signalized intersection, vehicles are cruising, decelerating, idling, and accelerating. Op-Mode distributions can be calculated from the ratios of individual mode travel times to total travel times on approach links and departure links. This type of information could be obtained from Op-Mode distribution data from (1) existing intersections with similar geometric and operational (traffic) characteristics, or (2) output from traffic simulation models for the proposed project or similar projects. Acceleration and deceleration assumptions, methods, and data underlying the activity-to-Op-Mode calculations should be documented as part of the PM hot-spot analysis.

The following methodology describes a series of equations to assist in calculating vehicle travel times on approach and departure links. Note that a single approach and single departure link should be defined to characterize vehicles approaching, idling at, and departing an intersection (e.g., there is no need for an "idling link," as vehicle idling is captured as part of the approach link).

D.4.1 Approach links

When modeling each approach link, the fraction of fleet travel times in seconds (s) in each mode of operation should be determined based on the fraction of time spent cruising, decelerating, accelerating, and idling:

Total Fleet Travel Time (s) = Cruise Time + Decel Time + Accel Time + Idle Time

The cruise travel time can be represented by the number of vehicles cruising multiplied by the length of approach divided by the average cruise speed:

Cruise Time (s) = Number of Cruising Vehicles * (Length of Approach (mi) ÷ Average Cruise Speed (mi/hr)) * 3600 s/hr

The deceleration travel time can be represented by the number of vehicles decelerating multiplied by the average cruise speed divided by the average deceleration rate:

Decel Time (s) = Number of Decelerating Vehicles * (Average Cruise Speed (mi/hr) ÷ Average Decel Rate (mi/hr/s))

The acceleration travel time occurring on an approach link can be similarly represented. However, to avoid double-counting acceleration activity that occurs on the departure link, users should multiply the acceleration time by the proportion of acceleration that occurs on the approach link (Accel Length Fraction on Approach):

Accel Time (s) = Number of Accelerating Vehicles * (Average Cruise Speed (mi/hr) ÷ Average Accel Rate (mi/hr/s)) * Accel Length Fraction on Approach

The idle travel time can be represented by the number of vehicles idling multiplied by the average stopped delay (average time spent stopped at an intersection):

Idle Time (s) = Number of Idling Vehicles * Average Stopped Delay (s)

Control delay (total delay caused by an intersection) may be used in lieu of average stopped delay, but control delay includes decelerating and accelerating travel times, which should be subtracted out (leaving only idle time).

After calculating the fraction of time spent in each mode of approach activity, users should select the appropriate MOVES Op-Mode corresponding to each particular type of activity (see Section 4.5.7 for more information). The operating modes in MOVES typifying approach links include:
- Cruise/acceleration (OpModeID 11-16, 22-25, 27-30, 33, 35, 37-40);
- Low and moderate speed coasting (OpModeID 11, 21);
- Braking (OpModeID 0, 501);

- Idling (OpModeID 1); and
- Tire wear (OpModeID 400-416).

The relative fleet travel time fractions can be allocated to the appropriate Op-Modes in MOVES. The resulting single Op-Mode distribution accounts for relative times spent in the different driving modes (cruise, deceleration, acceleration, and idle) for the approach link. A simple example of deriving Op-Mode distributions for a link using this methodology is demonstrated in Step 3 of Appendix F for a bus terminal facility.

D.4.2 Departure links

When modeling each departure link, the fraction of fleet travel times spent in each mode of operation should be determined based on the fraction of time spent cruising and accelerating:

Total Fleet Travel Time (s) = Cruise Time + Accel Time

The cruise travel time can be represented by the number of vehicles cruising multiplied by the travel distance divided by the average cruise speed:

Cruise Time (s) = Number of Cruising Vehicles * (Length of Departure (mi) ÷ Average Cruise Speed (mi/hr)) * 3600 s/hr

The acceleration travel time occurring during the departure link can be represented by the number of vehicles accelerating multiplied by the average cruise speed divided by the average acceleration rate. However, to avoid double-counting acceleration activity that occurs on the approach link, users should multiply the resulting acceleration time by the proportion of acceleration that occurs on the departure link (Accel Length Fraction on Departure):

Accel Time (s) = Number of Accelerating Vehicles * (Average Cruise Speed (mi/hr) ÷ Average Accel Rate (mi/hr/s)) * Accel Length Fraction on Departure

After calculating fraction of time spent in each mode of departure activity, users should select the appropriate MOVES Op-Mode corresponding to each particular type of activity (see Section 4.5.7 for more information). The operating modes typifying departure links include:
- Cruise/acceleration (OpModeID 11-16, 22-25, 27-30, 33, 35, 37-40); and
- Tire wear (OpModeID 401-416).

The relative fleet travel time fractions can be allocated to the appropriate Op-Modes. The resulting single Op-Mode distribution accounts for relative times spent in the different driving modes (cruise and acceleration) for the departure link.

Appendix E:
Example Quantitative PM Hot-spot Analysis of a Highway Project using MOVES and CAL3QHCR

E.1 INTRODUCTION

The purpose of this appendix is to demonstrate the procedures for completing a hot-spot analysis using MOVES and CAL3QHCR following the basic steps described in Section 3. Readers should reference the appropriate sections in the guidance as needed for more detail on how to complete each step of the analysis. This example is limited to showing the build scenario; in practice, project sponsors may also have to analyze the no-build scenario. While this example calculates emission rates using MOVES, EMFAC users may find the air quality modeling described in this appendix helpful.

Note: The following example of a quantitative PM hot-spot analysis is highly simplified and intended only to demonstrate the basic procedures described in the guidance. This example uses default data in places where the use of project-specific data in a real-world situation would be expected. In addition, actual PM hot-spot analyses could be significantly more complex and are likely to require more documentation of data and decisions. In this example, the interagency consultation process is used as needed for evaluating and choosing models, methods, and assumptions, according to the requirements of 40 CFR 93.105(c)(1)(i).

E.2 PROJECT DESCRIPTION AND CONTEXT

The proposed project is the construction of a highway interchange connecting a four-lane principle arterial with a six-lane freeway through on-and-off ramps (see Exhibit E-1). The project is being built to allow truck access to local businesses. The project is located in an area that was designated nonattainment for the 2006 24-hour $PM_{2.5}$ NAAQS and 1997 annual $PM_{2.5}$ NAAQS.

The following is some additional pertinent data about the project:
- The project is located in a medium-sized city (within one county) in a state other than California.
- The project is expected to take less than a year to complete and has an estimated completion date of 2013. The year of peak emissions is expected to be 2015, when considering the project's emissions and background concentrations.
- In 2015, the average annual daily traffic (AADT) at this location is expected to exceed 125,000 vehicles and greater than eight percent of the traffic will be heavy-duty diesel trucks.
- The area surrounding the proposed project is primarily residential, with no nearby sources that need to be included in air quality modeling.

- The state does not have an adequate or approved SIP budget for either $PM_{2.5}$ NAAQS, and neither the EPA nor the state air agency has made a finding that road dust is a significant contributor to the $PM_{2.5}$ nonattainment problem.

Exhibit E-1. Simple Diagram of the Proposed Highway Project

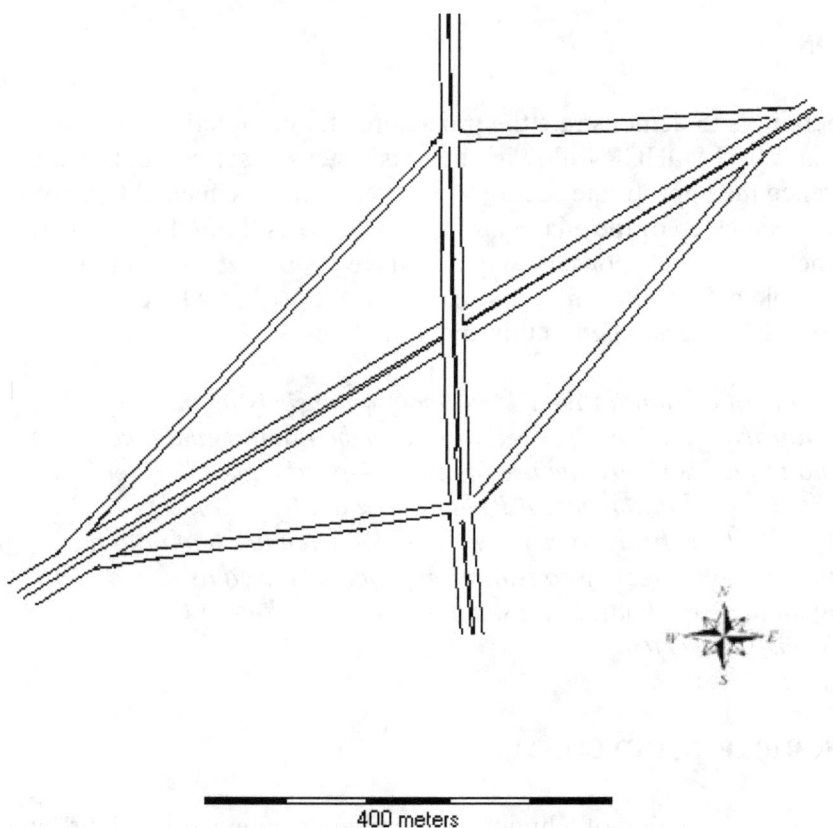

400 meters

E.3 DETERMINE NEED FOR PM HOT-SPOT ANALYSIS (STEP 1)

The proposed project is determined to be of local air quality concern under the conformity rule because it is a new freeway project with a significant number of diesel vehicles (see 40 CFR 93.123(b)(1)(i) and Sections 2.2 and Appendix B of the guidance). Therefore, a quantitative PM hot-spot analysis is required.

E.4 DETERMINE APPROACH, MODELS, AND DATA (STEP 2)

E.4.1 Determining geographic area and emission sources to be covered by the analysis

First, the interagency consultation process is used to ensure that the project area is defined so that the analysis includes the entire project, as required by 40 CFR 93.123(c)(2). As previously noted, it is also determined that, in this case, there are no nearby emission sources to be included in air quality modeling (see Section 3.3.2).

E.4.2 Deciding the general analysis approach and analysis year(s)

Second, the project sponsor determines that the preferred approach in this case is to model the build scenario first, completing a no-build scenario only if necessary.

In addition, it is determined that the year of peak emissions (within the timeframe of the current transportation plan) is mostly likely to be 2015. Therefore, 2015 is selected as the year of the analysis, and the analysis considers traffic data from 2015 (see Section 3.3.3).

E.4.3 Determining the PM NAAQS to be evaluated

Because the area has been designated nonattainment for both the 2006 24-hour $PM_{2.5}$ NAAQS and 1997 annual $PM_{2.5}$ NAAQS, the results of the analysis will have to be compared to both NAAQS (see Section 3.3.4). All four quarters are included in the analysis in order to estimate a year's worth of emissions for both NAAQS.

E.4.4 Deciding on the type of PM emissions to be modeled

Next, the following directly-emitted $PM_{2.5}$ emissions are determined to be relevant for estimating the emissions in the analysis (see Section 2.5):
- Vehicle exhaust[1]
- Brake wear
- Tire wear

E.4.5 Determining the models and methods to be used

Since this project is located outside of California, MOVES is used for emissions modeling. In addition, it is determined that, since this is a highway project with no nearby sources that need to be included in the air quality modeling, either AERMOD or CAL3QHCR could be used for air quality modeling (see Section 3.3.6). In this case, CAL3QHCR is selected. Making the decision on what air quality model to use at this stage is important so that the appropriate data are collected, among other reasons (see next step).

---

[1] Represented in MOVES as $PM_{total\ running}$ and $PM_{total\ crankcase\ running}$.

E.4.6   Obtaining project-specific data

Finally, the project sponsor compiles the data required to use MOVES, including project traffic data, vehicle types and age, and temperature and humidity data for the months and hours to be modeled (specifics on the data collected are described in the following steps). In addition, information necessary to use CAL3QHCR to model air quality is gathered, including meteorological data and information on representative air quality monitors. The sponsor also ensures the latest planning assumptions are used and that data used for the analysis are consistent with that used in the latest regional emissions analysis, as required by the conformity rule (see Section 3.3.7).

E.5   ESTIMATE ON-ROAD MOTOR VEHICLE EMISSIONS (STEP 3)

Having completed the analysis preparations described above, the project sponsor then follows the instructions provided in Section 4 of the guidance to use MOVES to estimate the project's on-road emissions:

E.5.1   Characterizing the project in terms of links

As described in Section 4.2 of the guidance, links are defined based on the expected emission rate variability across the project. Generally, a highway project like the one proposed in this example can be broken into four unique activity modes:
- Freeway driving at 55 mph;
- Arterial cruise at 45 mph;
- Acceleration away from intersections to a cruising speed of 45/55 mph; and
- Cruise, deceleration, and idle/cruise (depending on light timing) at intersections.

Following the guidance given in Section 4.2, 20 links are defined for MOVES and CAL3QHCR modeling, each representing unique geographic and activity parameters (see Exhibits E-2 and E-3). Each LinkID is defined with the necessary information for air quality modeling: link length, link width, link volume, as well as link start and end points (x1, y1, x2, y2 coordinates).

Decisions on how to best define links are based on an analysis of vehicle activity and patterns within the project area. AADT is calculated from a travel demand model for passenger cars, passenger trucks, intercity buses, short haul trucks, and long haul trucks. From these values, both an average-hour and peak-hour volume is calculated. The average and peak-hour vehicle counts for each part of the project are shown in Exhibit E-3.

Exhibit E-2. Diagram of Proposed Highway Project Showing Links

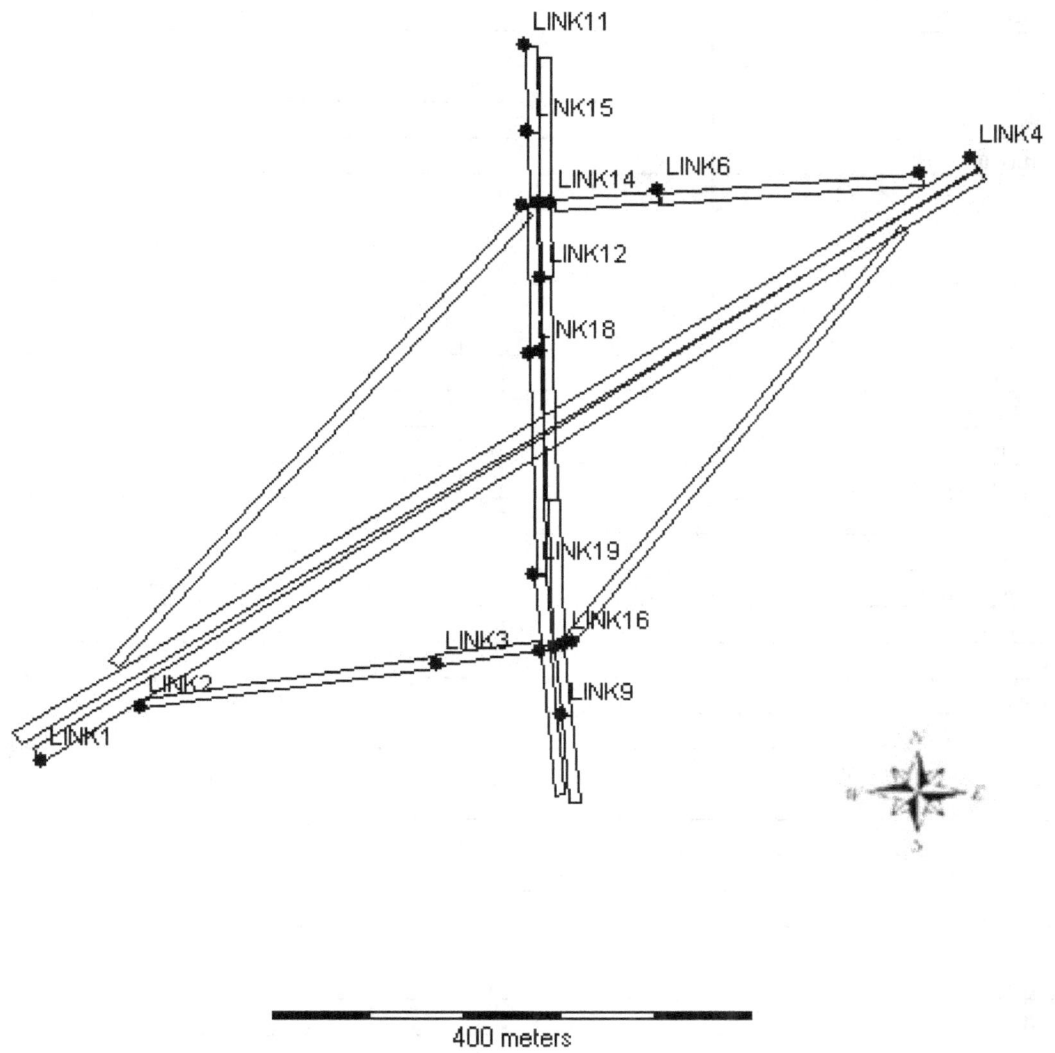

400 meters

Based on the conditions in the project area, for this analysis peak traffic is assumed to be representative of morning rush hour (AM: 6 a.m. to 9 a.m.) and evening rush hour (PM: 4 p.m. to 7 p.m.), while average hour traffic represents all other hours: midday (MD: 9 a.m. to 4 p.m.), and overnight (ON: 7 p.m. to 6 a.m.) Identical traffic volume and speed profiles are assumed for all quarters of the year. Quarters are defined as described in Section 3.3.4 of the guidance: Q1 (January-March), Q2 (April-June), Q3 (July-September), and Q4 (October-December).

Exhibit E-3. Peak-Hour and Average-Hour Traffic Counts for Each Project Link

Freeway	Peak Hour Count	Average Hour Count	Fraction of Total
Passenger Cars	2260	452	0.45
Passenger Trucks	1760	352	0.35
Intercity Buses	36	7	0.01
Short Haul Trucks (gas)	60	12	0.01
Long Haul Trucks (diesel)	944	189	0.19
Total	5060	1012	1.00

Exit Ramps	Peak Hour Count	Average Hour Count	Fraction of Total
Passenger Cars	124	25	0.22
Passenger Trucks	124	25	0.22
Intercity Buses	8	2	0.01
Short Haul Trucks (gas)	12	2	0.02
Long Haul Trucks (diesel)	300	60	0.53
Total	568	114	1.00

Entrance Ramps	Peak Hour Count	Average Hour Count	Fraction of Total
Passenger Cars	176	35	0.29
Passenger Trucks	148	30	0.24
Intercity Buses	0	0	0.00
Short Haul Trucks (gas)	16	3	0.03
Long Haul Trucks (diesel)	276	55	0.45
Total	616	123	1.00

Arterial Road	Peak Hour Count	Average Hour Count	Fraction of Total
Passenger Cars	124	25	0.22
Passenger Trucks	116	23	0.20
Intercity Buses	12	2	0.02
Short Haul Trucks (gas)	0	0	0.00
Long Haul Trucks (diesel)	316	63	0.56
Total	568	114	1.00

A significant amount of traffic using the project is expected to be diesel trucks. While the freeway contains approximately 19% diesel truck traffic, traffic modeling for the on- and off-ramps connecting the freeway to the arterial road suggests approximately half of vehicles are long-haul diesel trucks.

The average speeds on the freeway, arterial, and on/off-ramps are anticipated to be identical in the analysis year for both peak and average hours and assumed to approximately reflect the speed limit (55 mph, 45 mph, and 45 mph, respectively). Traffic flow through the two intersections north and south of the freeway is controlled by a signalized light with a 60% idle time for vehicles exiting the freeway and 40% idle time for traffic entering the freeway from the arterial road or traveling north and south on the arterial road passing over the freeway. The total project emissions, therefore, are determined to be a function of:

- Vehicles traveling east and west on the freeway at a relatively constant 55 mph;
- Exiting vehicles decelerating to a stop at either the north or south signalized intersection (or continuing through if the light is green);
- Vehicles accelerating away from the signalized intersections north and south, as well as accelerating to a 55 mph cruise speed on the on-ramps;
- Idling activity at both intersections during the red phase of the traffic light; and
- Vehicles traveling between the north and south intersections at a constant 45 mph.

As there is no new parking associated with the project (e.g., parking lots), there are no start emissions to be considered. Additionally, there are no trucks parked or "hoteling" in extended idle mode anywhere in the project area, so extended idle emissions do not need to be calculated.

E.5.2 Deciding how to handle link activity

As discussed in Section 4.2 of the guidance, MOVES offers several options for users to apply activity information to each LinkID. For illustrative purposes, based on the available information for the project (in this case, average speed, link average and peak volume, and red-light idle time) several methods of deriving Op-Mode distributions are employed in this example, as described below.

The links parameter table in Exhibit E-4 shows the various methods that activity is entered into MOVES for each link. The column "MOVES activity input" describes how the Op-Mode distribution is calculated for each particular link:

- Freeway links (links 1 and 4) are defined through a 55 mph average speed input, from which MOVES calculated an Op-Mode distribution (as described in Appendix D.2).
- Arterial cruise links (links 12 and 18) and links approaching an intersection queue (links 2, 5, 9 and 15) are defined through a link-drive schedule with a constant speed of 45 mph; indicating vehicles are cruising at 45 mph, with no acceleration or deceleration (as described in Appendix D.3).
- Links representing vehicles accelerating away from intersections (links 7, 8, 11, 14, 17, 20) are given "adjusted average speeds" calculated from guidance in the *Highway Capacity Manual 2000*, based on the link cruise speed (45 mph or 55 mph), red-light timing, and expected volume to capacity ratios. The adjusted average speeds (16.6 mph or 30.3 mph) are entered into MOVES, which calculates an Op-Mode distribution to reflect the lower average speed and subsequent higher emissions (as described in Appendix D.2).
- Queue links (links 3, 6, 10, 13, 16, and 19) are given an Op-Mode distribution that represents vehicles decelerating and idling (red light) as well as cruising through (green light) (as described in Appendix D.4).
  Step 1. First, an Op-Mode distribution is calculated for the link average speed (45 mph).

Step 2. Because this does not adequately account for idling at the intersection, the Op-Mode fractions are re-allocated to add in idling. For instance, after consulting the *Highway Capacity Manual 2000*, for this project scenario the red light timing corresponds to approximately 40% idle time. A fraction of 0.4 for Op-Mode "1" is therefore added to the Op-Mode distribution calculated from the 45 mph average speed in Step 1. The resulting Op-Mode distribution represents all activity on a queuing intersection link.

The length of the queue links are estimated as a function of the length of three trucks, one car, and one passenger truck with two meters in between each car and five meters in between each truck.

Departure links on the arterial road are assumed to have a link length of 125 meters (estimated to be the approximate distance that vehicles accelerate to a 45 mph cruising speed). The departure links from the intersection to the on-ramp are assumed to have a link length of 200 meters (estimated to be the approximate distance that vehicles accelerate to a 55 mph cruising speed).

Exhibit E-4. Link Parameters (Peak Traffic)

linkID	linkLength	linkwidth	linkVolume	linkAvgSpeed	adj. average speed	linkDescription	MOVES activity input	x1	y1	x2	y2
1	935	12	5060	55	n/a	EB highway	average speed	-422	-469	367	32
2	250	9	568	45	n/a	EB off-ramp cruise	**linkdrive schedule**	-337	-424	-89	-386
3	87	9	568	45	n/a	EB off-ramp queue	avg spd/opMode	-89	-386	-3	-372
4	940	12	5060	55	n/a	WB highway	average speed	358	44	-440	-453
5	220	9	568	45	n/a	WB off-ramp cruise	**linkdrive schedule**	315	29	96	13
6	87	9	568	45	n/a	WB off-ramp queue	avg spd/opMode	96	13	10	7
7	450	9	616	45	16.6	EB on-ramp	adj. average speed	19	-367	300	-13
8	520	9	616	45	16.6	WB on-ramp	adj. average speed	-14	2	-360	-386
9	75	9	568	45	n/a	sNB cruise	**linkdrive schedule**	26	-507	18	-433
10	61	9	568	45	n/a	sNB queue	avg spd/opMode	18	-433	12	-371
11	125	9	568	45	30.3	sNB depart	adj. average speed	12	-371	9	-246
12	190	9	568	45	n/a	NB connect	**linkdrive schedule**	9	-246	2	-56
13	61	9	568	45	n/a	nNB queue	avg spd/opMode	2	-56	1	5
14	125	9	568	45	30.3	nNB depart	adj. average speed	1	5	1	130
15	75	9	568	45	n/a	nSB cruise	**linkdrive schedule**	-10	142	-8	68
16	61	9	568	45	n/a	nSB queue	avg spd/opMode	-8	68	-9	6
17	125	9	568	45	30.3	nSB depart	adj. average speed	-9	6	-7	-122
18	189	9	568	45	n/a	SB connect	**linkdrive schedule**	-7	-122	-3	-311
19	61	9	568	45	n/a	sSB queue	avg spd/opMode	-3	-311	2	-371
20	125	9	568	45	30.3	sSB depart	adj. average speed	2	-371	12	-501

E.5.3   Determining the number of MOVES runs

Following the guidance given in Section 4.3, it is determined that 16 MOVES runs should be completed to produce emission factors that show variation across four hourly periods (12 a.m., 6 a.m., 12 p.m., and 6 p.m., corresponding to overnight, morning, midday, and evening traffic scenarios, respectively) and four quarterly periods (represented by the months of January, April, July, and October; see Section 3.3). MOVES will calculate values for all project links for the time period specified in each run. The 16 emission factors produced for each link are calculated as grams/vehicle-mile, which will then be paired with corresponding traffic volumes (peak or average hour, depending on the hour) and used in CAL3QHCR.

E.5.4   Developing basic run specification inputs

When configuring MOVES for the analysis, the project sponsor follows Section 4.4 of the guidance, including, but not limited to, the following:
- From the Scale panel, selecting the "Project" domain; in addition, choosing output in "Emission Rates," so that emission factors will be in grams/vehicle-mile as needed for CAL3QHCR (see Section 4.4.2).
- From the Time Spans panel, the appropriate year, month, day, and hour for each run is selected (see Section 4.4.3).
- From the Geographic Bounds panel, the custom domain is selected (see Section 4.4.4).
- From the Vehicles/Equipment panel, appropriate Source Types are selected (see Section 4.4.5).
- From the Road Types panel, Urban Restricted and Urban Unrestricted road types are selected (see Section 4.4.6).
- From the Pollutants and Processes panel, the appropriate pollutant/processes are selected according to Section 4.4.7 of the guidance for "highway links."
- In the Output panel, an output database is specified with grams and miles selected as units (see Section 4.4.10).

E.5.5   Entering project details using the Project Data Manager

Meteorology

As described previously, it is determined that MOVES should be run 16 times to reflect the following scenarios: 12 a.m., 6 a.m., 12 p.m., and 6 p.m. (corresponding to overnight, morning, midday, and evening traffic scenarios, respectfully) for the months of January, April, July, and October. Temperature and humidity data from a representative meteorological monitoring station are obtained and confirmed to be consistent with data used in the regional emissions analysis from the currently conforming transportation plan and TIP (see Section 4.5.1). Average values for each hour and month combination are used for each of the 16 MOVES runs. As an example, temperature and humidity values for 12 a.m. January are shown in Exhibit E-5.

Exhibit E-5. Temperature and Humidity Input (January 12 a.m.)

	A	B	C	D	E	F
	metjan12am.xls					
	monthID	zoneID	hourID	temperatur	relHumidity	
1						
2	1	990010	1	26.2	75.4	
3						
4						
5						
6						
7						

ZoneMonthHour / HourOfAi

Age Distribution

Section 4.5.2 of the guidance specifies that default data should be used only if an alternative local dataset cannot be obtained and the regional conformity analysis relies on national defaults. However, for the sake of simplicity only, in this example the national default age distribution for 2015 is used for all vehicles and all runs (see Exhibit E-6).

Exhibit E-6. Age Distribution Table (Partial)

agedist.xls

	A	B	C	D	E	F
1	sourceTyp	yearID	ageID	ageFraction		
2	21	2015	0	0.0599		
3	21	2015	1	0.0609		
4	21	2015	2	0.0616		
5	21	2015	3	0.0622		
6	21	2015	4	0.0620		
7	21	2015	5	0.0579		
8	21	2015	6	0.0559		
9	21	2015	7	0.0556		
10	21	2015	8	0.0578		
11	21	2015	9	0.0584		
12	21	2015	10	0.0591		
13	21	2015	11	0.0558		
14	21	2015	12	0.0497		
15	21	2015	13	0.0461		
16	21	2015	14	0.0404		
17	21	2015	15	0.0339		
18	21	2015	16	0.0286		
19	21	2015	17	0.0215		
20	21	2015	18	0.0163		
21	21	2015	19	0.0125		
22	21	2015	20	0.0109		
23	21	2015	21	0.0089		

sourceTypeAgeDistribution /

Fuel Supply and Fuel Formulation

In this example, it is determined appropriate to use the default fuel supply and formulation (see Exhibits E-7 and E-8). The default fuel supply and formulation are imported for each respective quarter (January, April, July, and October) and used for the corresponding MOVES runs.

Exhibit E-7. Fuel Supply Table

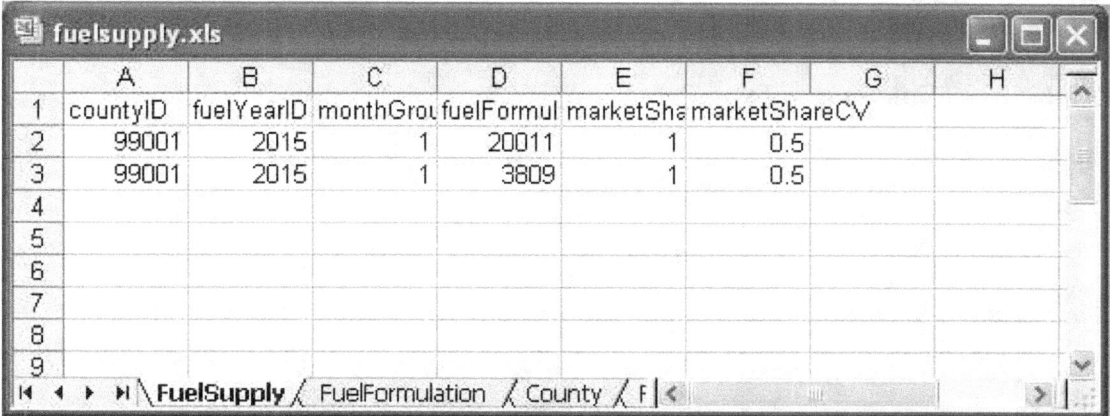

	A	B	C	D	E	F	G	H
1	countyID	fuelYearID	monthGrou	fuelFormul	marketSha	marketShareCV		
2	99001	2015	1	20011	1	0.5		
3	99001	2015	1	3809	1	0.5		
4								
5								
6								
7								
8								
9								

Exhibit E-8. Fuel Formulation Table

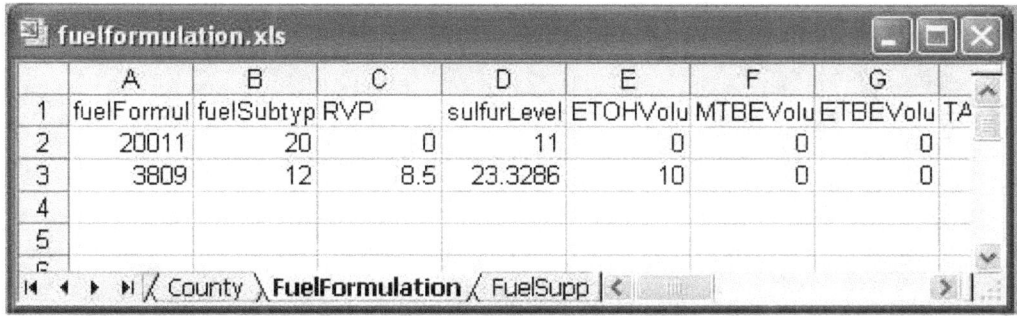

	A	B	C	D	E	F	G	
1	fuelFormul	fuelSubtyp	RVP	sulfurLevel	ETOHVolu	MTBEVolu	ETBEVolu	TA
2	20011	20	0	11	0	0	0	
3	3809	12	8.5	23.3286	10	0	0	
4								
5								

Inspection and Maintenance (I/M)

As there is no PM emissions benefit in MOVES for I/M programs, this menu item is skipped (see Section 4.5.4).

Link Source Type

The distribution of vehicle types on each link is defined in the Link Source Type table (Exhibit E-9) following the guidance in Section 4.5.5. The fractions are derived from the vehicle count estimates in Exhibit E-3.

Exhibit E-9. Link Source Type Table

	A	B	C	D	E
1	linkID	sourceTyp	sourceTypeHourFraction		
2	1	21	0.45		
3	1	31	0.35		
4	1	41	0.01		
5	1	61	0.01		
6	1	62	0.19		
7	2	21	0.22		
8	2	31	0.22		
9	2	41	0.01		
10	2	61	0.02		
11	2	62	0.53		
12	3	21	0.22		
13	3	31	0.22		
14	3	41	0.01		
15	3	61	0.02		
16	3	62	0.53		
17	4	21	0.45		
18	4	31	0.35		

linkSourceTypeHour / Sourc

Links

The Links input table shown in Exhibit E-10 is used to define each individual project link in MOVES. Road Types 4 and 5 indicate Urban Restricted (freeway) and Urban Unrestricted (arterial) road types, respectively; these correspond to the two road types represented in this example. The average speed is entered for all links, but only used to calculate Op-Mode distributions for links 1, 4, 7, 8, 11, 14, 17, and 20 (other links are explicitly defined with a link-drive schedule or Op-Mode distribution). Link length and link volume is entered for each link; however, since the "Emission Rates" option is selected in the Scale panel, MOVES will produce grams/vehicle-mile. The volume and link length will become relevant when running the air quality model later in this analysis.

Exhibit E-10. Links Input (AM Period)

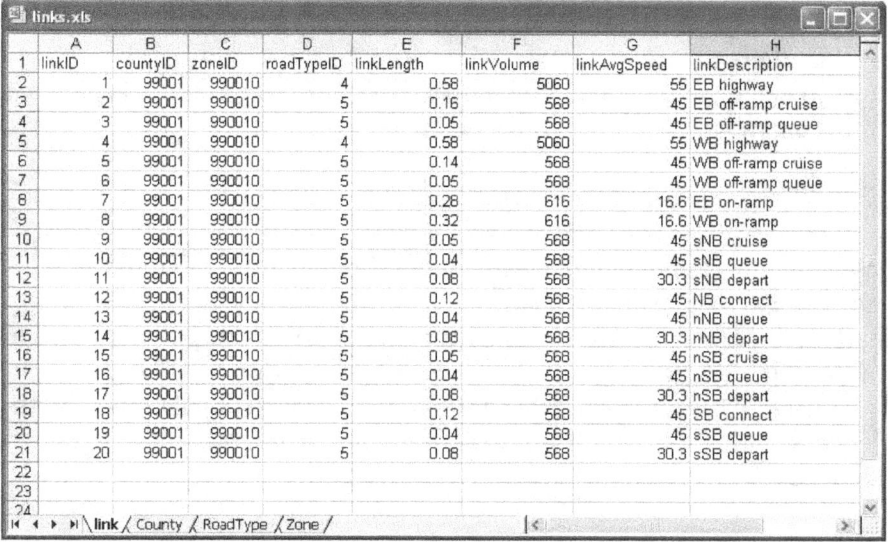

The remaining links are defined with an Op-Mode distribution (Exhibit E-11) calculated separately, as discussed earlier. Operating modes used in this analysis vary by both link and source type, but not by hour or day.

Exhibit E-11. Operating Mode Distribution Table (Partial)

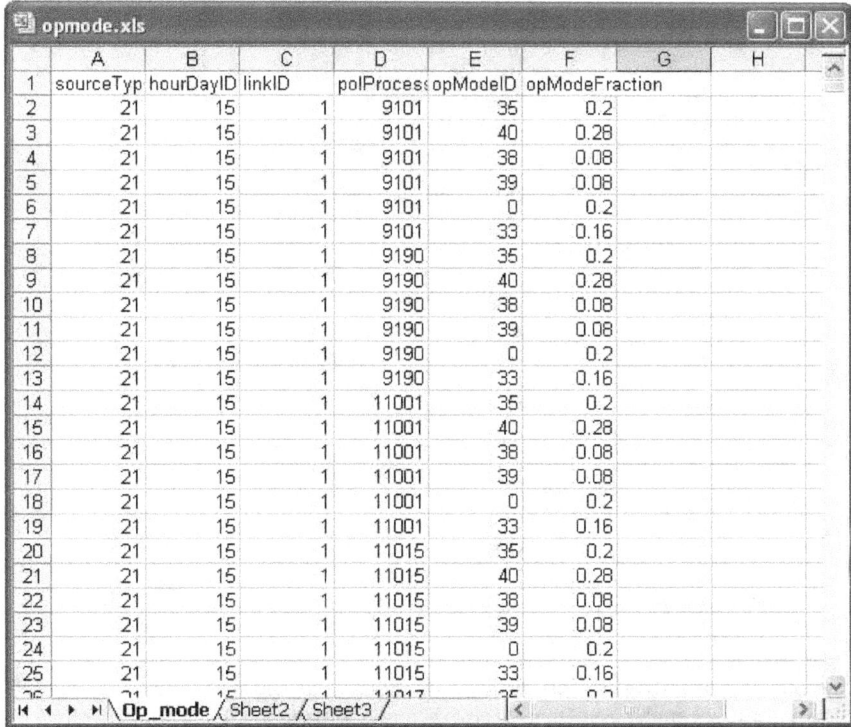

<u>Off-Network</u>

As it was determined that there are no off-network links (such as parking lots or truck stops) that would have to be considered using the Off-Network Importer, there is no need to use this option in this example.

E.5.6   Generating emission factors for use in air quality modeling

After generating the run specification and entering the required information into the Project Data Manager as described above, MOVES is run 16 times, once for each unique hour/month combination. Upon completion of each run, the MOVES output is located in the MySQL output database table "rateperdistance" and sorted by Month, Hour, LinkID, ProcessID, and PollutantID. An aggregate $PM_{2.5}$ emission factor is then calculated by the project sponsor for each Month, Hour, and LinkID combination using the following equation and the guidance given in Section 4.4.7 of the guidance:[2]

$$PM_{aggregate\ total} = (PM_{total\ running}) + (PM_{total\ crankcase\ running}) + (brake\ wear) + (tire\ wear)$$

The 16 resulting grams/vehicle-mile emission factors (Exhibit E-12) for each link are then ready to be used as input into the CAL3QHCR dispersion model to predict future $PM_{2.5}$ concentrations.

---

[2] EPA is considering creating one or more MOVES scripts that would automate the summing of aggregate emissions when completing project-level analyses. These scripts would be made available for download on the MOVES website (www.epa.gov/otaq/models/moves/tools.htm), when available.

Exhibit E-12. Grams/Vehicle-Mile Emission Factors Calculated from MOVES Output by Link, Quarter, and Hour

output_summary.xls

	A	B	C	D	E	F	G	H	I	J
1	linkID	linkLength (miles)	Jan12am	Jan6am	Jan12pm	Jan6pm	Apr12am	Apr6am	Apr12pm	Apr6pm
2	1	0.58	0.121	0.128	0.113	0.111	0.098	0.103	0.090	0.089
3	2	0.16	0.374	0.374	0.373	0.373	0.371	0.372	0.371	0.370
4	3	0.05	0.260	0.265	0.255	0.254	0.246	0.249	0.242	0.241
5	4	0.58	0.121	0.128	0.113	0.111	0.098	0.103	0.090	0.089
6	5	0.14	0.374	0.374	0.373	0.373	0.371	0.372	0.371	0.370
7	6	0.05	0.260	0.265	0.255	0.254	0.246	0.249	0.242	0.241
8	7	0.28	0.539	0.552	0.524	0.522	0.498	0.507	0.484	0.482
9	8	0.32	0.539	0.552	0.524	0.522	0.498	0.507	0.484	0.482
10	9	0.05	0.399	0.399	0.398	0.398	0.396	0.397	0.396	0.395
11	10	0.04	0.336	0.342	0.328	0.327	0.316	0.320	0.309	0.308
12	11	0.08	0.364	0.370	0.357	0.356	0.346	0.350	0.340	0.339
13	12	0.12	0.399	0.399	0.398	0.398	0.396	0.397	0.396	0.395
14	13	0.04	0.336	0.342	0.328	0.327	0.316	0.320	0.309	0.308
15	14	0.08	0.364	0.370	0.357	0.356	0.346	0.350	0.340	0.339
16	15	0.05	0.399	0.399	0.398	0.398	0.396	0.397	0.396	0.395
17	16	0.04	0.336	0.342	0.328	0.327	0.316	0.320	0.309	0.308
18	17	0.08	0.364	0.370	0.357	0.356	0.346	0.350	0.340	0.339
19	18	0.12	0.399	0.399	0.398	0.398	0.396	0.397	0.396	0.395
20	19	0.04	0.336	0.342	0.328	0.327	0.316	0.320	0.309	0.308
21	20	0.08	0.364	0.370	0.357	0.356	0.346	0.350	0.340	0.339
22	linkID	linkLength (miles)	Jul12am	Jul6am	Jul12pm	Jul6pm	Oct12am	Oct6am	Oct12pm	Oct6pm
23	1	0.58	0.085	0.086	0.084	0.084	0.096	0.099	0.088	0.088
24	2	0.16	0.369	0.369	0.369	0.369	0.370	0.371	0.370	0.370
25	3	0.05	0.238	0.239	0.237	0.237	0.244	0.247	0.240	0.240
26	4	0.58	0.085	0.086	0.084	0.084	0.096	0.099	0.088	0.088
27	5	0.14	0.369	0.369	0.369	0.369	0.370	0.371	0.370	0.370
28	6	0.05	0.238	0.239	0.237	0.237	0.244	0.247	0.240	0.240
29	7	0.28	0.469	0.472	0.468	0.468	0.489	0.495	0.475	0.476
30	8	0.32	0.469	0.472	0.468	0.468	0.489	0.495	0.475	0.476
31	9	0.05	0.394	0.394	0.394	0.394	0.395	0.396	0.394	0.395
32	10	0.04	0.304	0.305	0.303	0.303	0.313	0.316	0.306	0.307
33	11	0.08	0.332	0.333	0.331	0.331	0.340	0.343	0.334	0.334
34	12	0.12	0.394	0.394	0.394	0.394	0.395	0.396	0.394	0.395
35	13	0.04	0.304	0.305	0.303	0.303	0.313	0.316	0.306	0.307
36	14	0.08	0.332	0.333	0.331	0.331	0.340	0.343	0.334	0.334
37	15	0.05	0.394	0.394	0.394	0.394	0.395	0.396	0.394	0.395
38	16	0.04	0.304	0.305	0.303	0.303	0.313	0.316	0.306	0.307
39	17	0.08	0.332	0.333	0.331	0.331	0.340	0.343	0.334	0.334
40	18	0.12	0.394	0.394	0.394	0.394	0.395	0.396	0.394	0.395
41	19	0.04	0.304	0.305	0.303	0.303	0.313	0.316	0.306	0.307
42	20	0.08	0.332	0.333	0.331	0.331	0.340	0.343	0.334	0.334

H ◄ ► H \ Output \ **gramspervehiclemile** / CAL3QHCR Inputs /

E.6 ESTIMATE EMISSIONS FROM ROAD DUST, CONSTRUCTION, AND ADDITIONAL SOURCES (STEP 4)

E.6.1 Estimating re-entrained road dust

In this case, this area does not have any adequate or approved SIP budgets for either $PM_{2.5}$ NAAQS, and neither the EPA nor the state air agency have made a finding that road dust emissions are a significant contributor to the air quality problem for either $PM_{2.5}$ NAAQS. Therefore, $PM_{2.5}$ emissions from road dust do not need to be considered in this analysis (see Sections 2.5.3 and 6.2).

E.6.2 Estimating transportation-related construction dust

The construction of this project will not occur during the analysis year. Therefore, emissions from construction dust are not included in this analysis (see Sections 2.5.5 and 6.4).

E.6.3 Estimating additional sources of emissions in the project area

It is determined that the project area in the analysis year does not include locomotives or other nearby emission sources that have to be considered in the air quality modeling (see Section 6.6).

E.7 SELECT AN AIR QUALITY MODEL, DATA INPUTS, AND RECEPTORS (STEP 5)

E.7.1 Characterizing emission sources

As discussed previously, the CAL3QHCR model is selected to estimate $PM_{2.5}$ concentrations for this analysis (see Section 7.3). Each link is defined in CAL3QHCR with coordinates and dimensions matching the project parameters (shown in Exhibit E-4). The necessary inputs for link length, traffic volume, and corresponding link emission factor are also added using the CAL3QHCR Tier II approach. Each MOVES emission factor (12 a.m., 6 a.m., 12 p.m., and 6 p.m.) and traffic volume (average or peak) for each link is applied to multiple hours of the day, as follows:

- Morning peak (AM) emissions based on traffic data and meteorology occurring between 6 a.m. and 9 a.m.;
- Midday (MD) emissions based on data from 9 a.m. to 4 p.m.;
- Evening peak (PM) emissions based on data from 4 p.m. to 7 p.m.;
- Overnight (ON) emissions based on data from 7 p.m. to 6 a.m.

In addition, these factors are applied to each of the four quarters being modeled.

CAL3QHCR scenarios are built to model traffic conditions for all 24 hours of a weekday in each quarter (partial elements of the CAL3QHCR input file can be found in Exhibits E-13a and 13b, as file requires one to scroll down the screen): in all, four separate scenarios.

Exhibit E-13a. CAL3QHCR Quarter 1, 6 a.m. Input File (Partial)

```
highway_jan.INP - Notepad
File  Edit  Format  View  Help
'Hot-Spot Highway Example'  60. 175.   0.  0.   41  1 0
 1 1 98 12 31 98
 94823  98  94823 98
 1 1 'U'
'1'      -42.9    -20      1.8
'2'      -28.8    -3.1     1.8
'3'      -16.5    29       1.8
'4'      -16.5    48.8     1.8
'5'      29.6     31.8     1.8
'6'      12.7     -101     1.8
'7'      -14.6    -100.1   1.8
'8'      15.5     -152.9   1.8
'9'      -14.6    -154.7   1.8
'10'     -12.7    -220.7   1.8
'11'     17.4     -205.6   1.8
'12'     -11.8    -265.9   1.8
'13'     21.2     -257.4   1.8
'14'     19.3     -334.7   1.8
'15'     35.3     -333.7   1.8
'16'     34.4     -317.7   1.8
'17'     -21.2    -395.9   1.8
'18'     -23.1    -349.7   1.8
'19'     24       -18.1    1.8
'20'     -31.6    15.8     1.8
'21'     -3.3     -435.5   1.8
'22'     12.7     -7.8     1.8
'23'     24       -379     1.8
'24'     28.7     -411     1.8
'25'     45.7     -353.5   1.8
'26'     -9.9     -360.1   1.8
'27'     -12.7    -320.5   1.8
'28'     -42.9    -365.8   1.8
'29'     -19.3    -21      1.8
'30'     -22.2    -57.7    1.8
'31'     10.8     19.6     1.8
'32'     46.6     20.5     1.8
'33'     10.8     52.5     1.8
'34'     13.6     -32.3    1.8
'35'     48.5     -6.8     1.8
'36'     -7.1     -386.5   1.8
'37'     -43.8    -394     1.8
'38'     -6.2     -410     1.8
'39'     43.8     -387.4   1.8
'40'     -33.5    -38.9    1.8
'41'     33.4     -432.7   1.8
```

Exhibit E-13b. CAL3QHCR Quarter 1, 6 a.m. Input File (Partial)

```
highway_jan.INP - Notepad
File  Edit  Format  View  Help
2 'p'
1 1 1 1 1 1 1
'Example Highway Project'          20
   1  1
'EB highway'          'ag'   -422    367    -469    32      0      12
   2  1
'EB off-ramp cruise'  'ag'   -337    -89    -424    -386    0      9
   3  1
'EB off-ramp queue'   'ag'   -89     -3     -386    -372    0      9
   4  1
'WB highway'          'ag'   358     -440   44      -453    0      12
   5  1
'WB off-ramp cruise'  'ag'   315     96     29      13      0      9
   6  1
'WB off-ramp queue'   'ag'   96      10     13      7       0      9
   7  1
'EB on-ramp'          'ag'   19      300    -367    -13     0      9
   8  1
'WB on-ramp'          'ag'   -14     -360   2       -386    0      9
   9  1
'sNB cruise'          'ag'   26      18     -507    -433    0      9
  10  1
'sNB queue'           'ag'   18      12     -433    -371    0      9
  11  1
'sNB depart'          'ag'   12      9      -371    -246    0      9
  12  1
'NB connect'          'br'   9       2      -246    -56     0      9
  13  1
'nNB queue'           'ag'   2       1      -56     5       0      9
  14  1
'nNB depart'          'ag'   1       1      5       130     0      9
  15  1
'nSB cruise'          'ag'   -10     -8     142     68      0      9
  16  1
'nSB queue'           'ag'   -8      -9     68      6       0      9
  17  1
'nSB depart'          'ag'   -9      -7     6       -122    0      9
  18  1
'SB connect'          'br'   -7      -3     -122    -311    0      9
  19  1
'sSB queue'           'ag'   -3      2      -311    -371    0      9
  20  1
'sSB depart'          'ag'   2       12     -371    -501    0      9
01        0.0
          1       5060    0.1214
          2       568     0.3737
          3       568     0.2605
          4       5060    0.1214
          5       568     0.3737
          6       568     0.2605
          7       616     0.5392
          8       616     0.5392
          9       568     0.3985
          10      568     0.3356
          11      568     0.3642
```

Section 7.5 of the guidance recommends that users run the air quality model for five years of meteorological data when site-specific meteorology data is not available. Since CAL3QHCR can only process one year of meteorological data for each run, each quarterly scenario is run for five years of meteorological data for a total of 20 runs.[3]

---

[3] As explained in Section 7, AERMOD allows five years of meteorological data to be modeled in a single run (see Section 7.5.3)

E.7.2 Incorporating meteorological data

A representative set of meteorology data, as well as an appropriate surface roughness, are selected (see Section 7.5). The recommended five years of meteorological data are obtained from a local airport for calendar years 1998-2002. A surface roughness of 175 cm is selected for the site, consistent with the recommendations made in the Section 7.

E.7.3 Placing receptors

Using the guidance given in Section 7.6, receptors are placed at appropriate locations within the area substantially affected by the project (Exhibit E-14).[4] Note that this grid is shown for illustrative purposes only; placement, location, and spacing of actual receptors should follow the guidance in Section 7.6. Receptor heights are set at 1.8 meters. Additionally, a background concentration of "0" is input into the model. Representative background concentrations are added later (see Step 7).

CAL3QHCR is then run with five years of meteorological data (1998 through 2002) and output is produced for all receptors for each of the five years of meteorological data.

Exhibit E-14. Receptor Locations for Air Quality Modeling

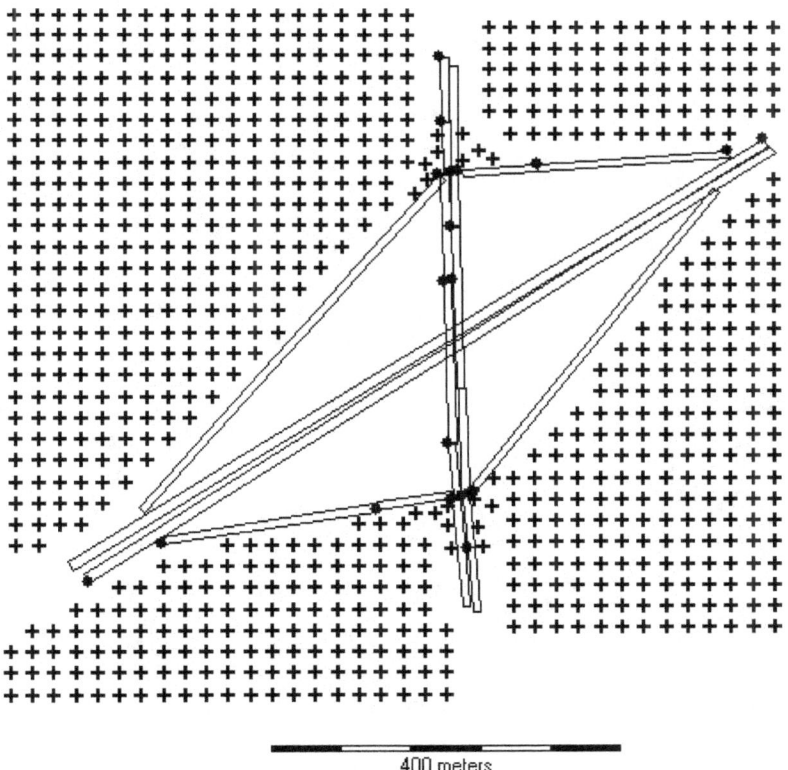

400 meters

---

[4] The number and arrangement of receptors used in this example are simplified for ease of explanation.

E.8 DETERMINE BACKGROUND CONCENTRATIONS FROM NEARBY AND OTHER EMISSION SOURCES (STEP 6)

Through the interagency consultation process, a nearby upwind $PM_{2.5}$ monitor that has been collecting ambient data for both the annual and 24-hour $PM_{2.5}$ NAAQS is determined to be representative of the background air quality at the project location. The most recent data set is used (in this case, calendar year 2008 through 2010) and average 24-hour $PM_{2.5}$ values are taken in a four-day/three-day measurement interval. As previously noted, no nearby sources needing to be included in the air quality model are identified.

Note: This is a highly simplified situation for illustrative purposes; refer to Section 8 of the guidance for additional considerations for how to most accurately reflect background concentrations in a real-world scenario.

E.9 CALCULATE DESIGN VALUES AND DETERMINE CONFORMITY (STEP 7)

With both CAL3QHCR outputs and background concentrations now available, the project sponsor can calculate the design values. For illustrative purposes, calculations for a single receptor with the highest modeled concentrations for the build scenario are shown in this example.[5] In this step, the guidance from Sections 9.3.2 and 9.3.3 are used to calculate design values from the modeled results and the background concentrations for comparison with the annual and 24-hour $PM_{2.5}$ NAAQS.

E.9.1 Determining conformity to the annual $PM_{2.5}$ NAAQS

First, average background concentrations are determined for each year of monitored data (shown in Exhibit E-15).

Exhibit E-15. Annual Average Background Concentration for Each Year

Monitoring Year	Annual Average Background Concentration
2008	13.348
2009	12.785
2010	13.927

The three-year average background concentration is then calculated (see Exhibit E-16).

Exhibit E-16. Calculation of Annual Design Value (At Highest Receptor)

Annual Average Background Concentration (Three-year Average)	Annual Average Modeled Concentration (Five-year Average)	Sum of Background + Project	Annual Design Value
13.353	1.580	14.933	14.9

---

[5] In an actual PM hot-spot analysis, design values would be calculated at additional receptors as described in Section 9.3.

To determine the annual $PM_{2.5}$ design value, the annual average background concentration is added to the five-year annual average modeled concentration (at the receptor with the highest annual average concentration from the CAL3QHCR output). This calculation is shown in Exhibit E-16. The sum (project + background) results in a design value of 14.9 $\mu g/m^3$. This value at the highest receptor is less than the 1997 annual $PM_{2.5}$ NAAQS of 15.0 $\mu g/m^3$. It can be assumed that all other receptors with lower modeled concentrations will also have design values less than this NAAQS. In this example it is unnecessary to determine appropriate receptors in the build scenario (per Section 9.4 of the guidance) or develop a no-build scenario for the annual $PM_{2.5}$ NAAQS, since the build scenario demonstrates that the hot-spot analysis requirements in the transportation conformity rule are met at all receptors.

E.9.2   Determining conformity to the 24-Hour $PM_{2.5}$ NAAQS

The next step is to calculate a design value to compare with the 2006 24-hour $PM_{2.5}$ NAAQS through a "Second Tier" analysis as described in Section 9.3.3. For ease of explanation, this process has been divided into individual steps, consistent with the guidance.

Step 7.1
The number of background measurements is counted for each year of monitored data (2008 to 2010). Based on a 4-day/3-day measurement interval, the dataset has 104 values per year.

Step 7.2
For each year of monitored concentrations, the eight highest daily background concentrations for each quarter are determined, resulting in 32 values (4 quarters; 8 concentrations/quarter) for each year of data (shown in Exhibit E-17).

Step 7.3
Identify the highest-predicted modeled concentration resulting from the project in each quarter, averaged across each year of meteorological data used for air quality modeling. For illustrative purposes, the highest average concentration across five years of meteorological data for a single receptor in each quarter is shown in Exhibit E-18. Note that, in a real-world situation, this process would be repeated for all receptors in the build scenario.

Exhibit E-17. Highest Daily Background Concentrations for Each Quarter and Each Year

2008				
Rank	Q1	Q2	Q3	Q4
1	20.574	21.262	22.354	20.434
2	20.152	20.823	22.042	20.016
3	19.743	20.398	21.735	19.611
4	19.346	19.985	21.434	19.218
5	18.961	19.584	21.140	18.837
6	18.588	19.196	20.851	18.467
7	18.226	18.819	20.568	18.109
8	17.874	18.454	20.291	17.761
2009				
Rank	Q1	Q2	Q3	Q4
1	20.195	20.867	21.932	20.058
2	19.784	20.440	21.628	19.651
3	19.386	20.026	21.329	19.257
4	19.000	19.624	21.037	18.875
5	18.625	19.235	20.750	18.504
6	18.262	18.857	20.469	18.145
7	17.910	18.490	20.194	17.796
8	17.568	18.135	19.924	17.457
2010				
Rank	Q1	Q2	Q3	Q4
1	21.137	21.847	22.980	20.990
2	20.698	21.390	22.655	20.556
3	20.272	20.948	22.336	20.135
4	19.860	20.519	22.023	19.726
5	19.459	20.102	21.717	19.330
6	19.071	19.698	21.417	18.945
7	18.694	19.307	21.123	18.572
8	18.329	18.927	20.834	18.211

Exhibit E-18. Five-year Average 24-hour Modeled Concentrations for Each Quarter (At Example Receptor)

	Q1	Q2	Q3	Q4
Five Year Average Maximum Concentration (At Example Receptor)	10.42	10.62	10.74	10.61

Step 7.4

The highest modeled concentration in each quarter (from Step 7.3) is added to each of the eight highest monitored concentrations for the same quarter for each year of monitoring data (from Step 7.2). As shown in Exhibit E-19, this step results in eight concentrations in each of four quarters for a total of 32 values for each year of monitoring data. As mentioned, this example analysis shows only a single receptor's values, but project sponsors may need to calculate design values at all receptors in the build scenario (see Section 9.3 of the guidance).

Exhibit E-19. Sum of Background and Modeled Concentrations at Example Receptor for Each Quarter

2008				
Rank	Q1	Q2	Q3	Q4
1	31.084	31.902	32.994	31.074
2	30.662	31.463	32.682	30.656
3	30.253	31.038	32.375	30.251
4	29.856	30.625	32.074	29.858
5	29.471	30.224	31.780	29.477
6	29.098	29.836	31.491	29.107
7	28.736	29.459	31.208	28.749
8	28.384	29.094	30.931	28.401

2009				
Rank	Q1	Q2	Q3	Q4
1	30.705	31.507	32.572	30.698
2	30.294	31.080	32.268	30.291
3	29.896	30.666	31.969	29.897
4	29.510	30.264	31.677	29.515
5	29.135	29.875	31.390	29.144
6	28.772	29.497	31.109	28.785
7	28.420	29.130	30.834	28.436
8	28.078	28.775	30.564	28.097

2010				
Rank	Q1	Q2	Q3	Q4
1	31.647	32.487	33.620	31.630
2	31.208	32.030	33.295	31.196
3	30.782	31.588	32.976	30.775
4	30.370	31.159	32.663	30.366
5	29.969	30.742	32.357	29.970
6	29.581	30.338	32.057	29.585
7	29.204	29.947	31.763	29.212
8	28.839	29.567	31.474	28.851

Step 7.5

As shown in Exhibit E-20, for each year of monitoring data, the 32 values from Step 7.4 are ordered together in a column and assigned a yearly rank for each value, from 1 (highest concentration) to 32 (lowest concentration).

Exhibit E-20. Ranking Sum of Background and Modeled Concentrations at Example Receptor for Each Year of Background Data

Rank	2008	2009	2010
1	32.994	32.572	33.620
2	32.682	32.268	33.295
3	**32.375**	**31.969**	**32.976**
4	32.074	31.677	32.663
5	31.902	31.507	32.487
6	31.780	31.390	32.357
7	31.491	31.109	32.057
8	31.463	31.080	32.030
9	31.208	30.834	31.763
10	31.084	30.705	31.647
11	31.074	30.698	31.630
12	31.038	30.666	31.588
13	30.931	30.564	31.474
14	30.662	30.294	31.208
15	30.656	30.291	31.196
16	30.625	30.264	31.159
17	30.253	29.897	30.782
18	30.251	29.896	30.775
19	30.224	29.875	30.742
20	29.858	29.515	30.370
21	29.856	29.510	30.366
22	29.836	29.497	30.338
23	29.477	29.144	29.970
24	29.471	29.135	29.969
25	29.459	29.130	29.947
26	29.107	28.785	29.585
27	29.098	28.775	29.581
28	29.094	28.772	29.567
29	28.749	28.436	29.212
30	28.736	28.420	29.204
31	28.401	28.097	28.851
32	28.384	28.078	28.839

Step 7.6
For each year of monitoring data, the value with a rank that corresponds to the projected 98th percentile concentration is determined. As discussed in Section 9, an analysis employing 101-150 background values for each year (as noted in Step 7.1, this analysis uses 104 values per year) uses the 3rd highest rank to represent a 98th percentile. The 3rd highest concentration (highlighted in Exhibit E-20) is referred to as the "projected 98th percentile concentration."

Step 7.7
Steps 7.1 through 7.6 are repeated to calculate a projected 98th percentile concentration at each receptor based on each year of monitoring data and modeled concentrations.

Step 7.8
For the example receptor, the average of the three projected 98th percentile concentrations (see Step 7.6) is calculated.

Step 7.9
The resulting value of 32.440 $\mu g/m^3$ is then rounded to the nearest whole $\mu g/m^3$, resulting in a design value at the example receptor of 32 $\mu g/m^3$. At each receptor this process should be repeated. In the case of this analysis, the example receptor is the receptor with the highest design value in the build scenario.

Step 7.10
The design values calculated at each receptor are compared to the NAAQS. In the case of this example, the highest 24-hour design value (32 $\mu g/m^3$) is less than the 2006 24-hour $PM_{2.5}$ NAAQS of 35 $\mu g/m^3$. Since this is the design value at the highest receptor, it can be assumed that the conformity requirements are met at all receptors in the build scenario. Therefore, it is unnecessary for the project sponsor to calculate design values for the no-build scenario for the 24-hour NAAQS.

E.10 CONSIDER MITIGATION AND CONTROL MEASURES (STEP 8)

In this case, the project is determined to conform. In situations when this is not the case, it may be necessary to consider additional mitigation or control measures. If measures are considered, additional air quality modeling would need to be completed and new design values calculated to ensure that conformity requirements are met. See Section 10 for more information, including some specific measures that might be considered.

E.11 DOCUMENT THE PM HOT-SPOT ANALYSIS (STEP 9)

The final step is to properly document the PM hot-spot analysis in the conformity determination (see Section 3.10).

Appendix F:
Example Quantitative PM Hot-spot Analysis of a Transit Project using MOVES and AERMOD

F.1   INTRODUCTION

The purpose of this appendix is to demonstrate the procedures for completing a hot-spot analysis using MOVES and AERMOD following the basic steps described in Section 3. Readers should reference the appropriate sections in the guidance as needed for more detail on how to complete each step of the analysis. This example is limited to showing the build scenario; in practice, project sponsors may have to also analyze the no-build scenario. While this example calculates emission rates using MOVES, EMFAC users may find the air quality modeling described in this appendix helpful.

Note: The following example of a quantitative PM hot-spot analysis is highly simplified and intended only to demonstrate the basic procedures described in the guidance. This example uses default data in places where the use of project-specific data in a real-world situation would be expected. In addition, actual PM hot-spot analyses could be significantly more complex and are likely to require more documentation of data and decisions. In this example, the interagency consultation process is used as needed for evaluating and choosing models, methods, and assumptions, according to the requirements of 40 CFR 93.105(c)(1)(i).

F.2   PROJECT DESCRIPTION AND CONTEXT

The proposed project is a new regionally significant bus terminal that would be created by taking a downtown street segment one block in length and reserving it for bus use only. It would be an open-air facility containing six "sawtooth" lanes where buses enter to load and unload passengers. The terminal is designed to handle about 575 diesel buses per day with up to 48 buses in the peak hour. The project is located in an area designated nonattainment for the 2006 24-hour $PM_{2.5}$ NAAQS and 1997 annual $PM_{2.5}$ NAAQS.

The following is some additional pertinent data about the project:
- The proposed project is located in a medium-size city (within one county) in a state other than California.
- The project is expected to take less than a year to complete and has an estimated completion date of 2013. The year of peak emissions is expected to be 2015, when considering the project's emissions and background concentrations.
- The area surrounding the proposed project is primarily commercial, with no nearby sources of $PM_{2.5}$ that need to be modeled. This assumption is made to simplify the example. In most cases, transit projects include parking lots with emissions that would be considered in a PM hot-spot analysis.

- The state does not have an adequate or approved SIP budget for either $PM_{2.5}$ NAAQS, and neither the EPA nor the state air quality agency has made a finding that road dust is a significant contributor to the $PM_{2.5}$ nonattainment problem.

F.3   DETERMINE NEED FOR PM HOT-SPOT ANALYSIS (STEP 1)

The proposed project is determined to be of local air quality concern under the conformity rule because it is a new bus terminal that has a significant number of diesel vehicles congregating at a single location (see 40 CFR 93.123(b)(1)(iii) and Section 2.2 and Appendix B of the guidance). Therefore, a quantitative PM hot-spot analysis is required.

F.4   DETERMINE APPROACH, MODELS, AND DATA (STEP 2)

F.4.1   *Determining geographic area and emission sources to be covered by the analysis*

First, the interagency consultation process is used to ensure that the project area is defined so that the analysis includes the entire project, as required by 40 CFR 93.123(c)(2). As previously noted, it is also determined that, in this case, there are no nearby emission sources to be modeled (see Section 3.3.2).

F.4.2   *Deciding the general analysis approach and analysis year(s)*

The project sponsor then determines that the preferred approach in this case is to model the build scenario first, completing a no-build scenario only if necessary.

The year of peak emissions (within the timeframe of the current transportation plan) is determined to be 2015. Therefore, 2015 is selected as the year of the analysis, and the analysis will consider traffic data from 2015 (see Section 3.3.3).

F.4.3   *Determining the PM NAAQS to be evaluated*

Because the area has been designated nonattainment for both the 2006 24-hour $PM_{2.5}$ NAAQS and 1997 annual $PM_{2.5}$ NAAQS, the results of the analysis will be compared to both NAAQS (see Section 3.3.4). All four quarters are included in the analysis in order to estimate a year's worth of emissions for both NAAQS.

F.4.4 Deciding on the type of PM emissions to be modeled

Next, the following directly-emitted PM emissions are determined to be relevant for estimating the emissions in the analysis (see Section 2.5):
- Vehicle exhaust[1]
- Brake wear
- Tire wear

F.4.5 Determining the models and methods to be used

Since this project will be located outside of California, MOVES is used for emissions modeling. In addition, it is determined that, since this is a terminal project, the appropriate air quality model to use would be AERMOD (see Section 3.3.6). Making the decision on what air quality model to use at this stage is important so that the appropriate data are collected, among other reasons (see next step).

F.4.6 Obtaining project-specific data

Finally, having selected a model and a general modeling approach, the project sponsor compiles the data required to use MOVES, including project traffic data, vehicle types and age, and temperature and humidity data for the months and hours to be modeled (specifics on the data collected are described in the following steps). In addition, information required to use AERMOD to model air quality is gathered, including meteorological data and information on representative air quality monitors. The sponsor ensures the latest planning assumptions are used and that data used for the analysis are consistent with that used in the latest regional emissions analysis, as required by the conformity rule (see Section 3.3.7).

F.5 ESTIMATE ON-ROAD MOTOR VEHICLE EMISSIONS (STEP 3)

Having completed the analysis preparations described above, the project sponsor then follows the instructions provided in Section 4 of the guidance to use MOVES to estimate the on-road emissions from this terminal project:

F.5.1 Characterizing the project in terms of links

Using the guidance described in Section 4.2, a series of links are defined in order to accurately capture the activity at the proposed terminal. As shown in Exhibit F-1, two one-way running links north and south of the facility ("Link 1" and "Link 2") are defined to describe buses entering and exiting the terminal. A third running/idle link (shown as "Link 3" to the north of the facility), is used to describe vehicles idling at the signalized

---

[1] Represented in MOVES as $PM_{total\ running}$, $PM_{total\ crankcase\ running}$, $PM_{total\ ext.\ idle}$, and $PM_{total\ crankcase\ ext.\ idle}$.

light before exiting the facility. Links 4 through 9 represented bus bays where buses drop-off and pick-up passengers; these are referred to as the terminal links.[2]

Exhibit F-1. Diagram of Proposed Bus Terminal Showing Links

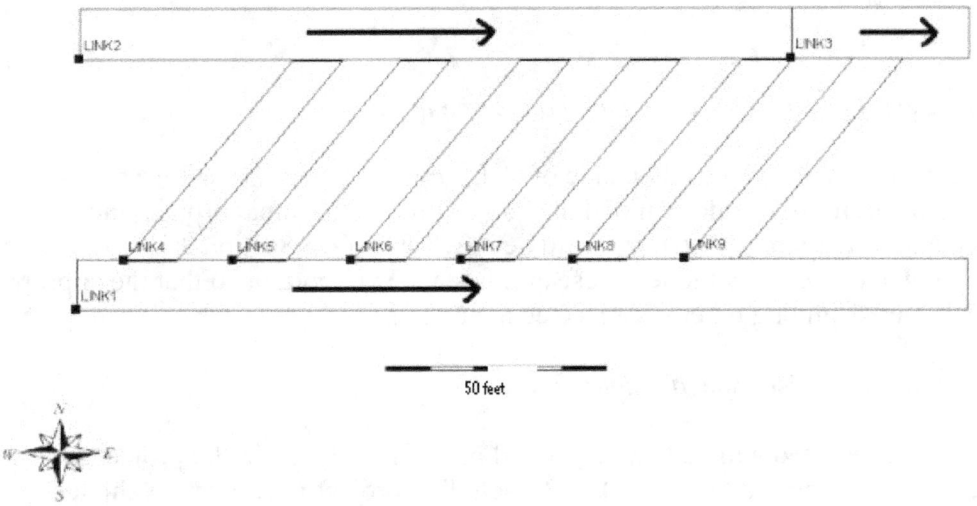

The running links have the following dimensions:
  Link 1: 200 feet long by 24 feet wide
  Link 2: 160 feet long by 24 feet wide
  Link 3: 40 feet long by 24 feet wide

Additionally, the dimensions of the six terminal links (Links 4 through 9) are 60 feet long by 12 feet wide. These links are oriented diagonally from southwest to northeast. The queue link (Link 3) is defined with a length of 40 feet, based on the average length of a transit bus.

After identifying and defining the links, traffic conditions are estimated for the project in the analysis year of 2015. The terminal was presumed to be in operation all hours of the year. Based on expected terminal operations, the anticipated future traffic volumes are available for each hour of an average weekday (see Exhibit F-2). To simplify the analysis, the sponsor conservatively assumes weekday traffic for all days of the year, even though the operating plan calls for reduced service on weekends.[3] Identical traffic volume and activity profiles are assumed for all quarters of the year. Quarters are defined for this analysis as described in Section 3.3.4 of the guidance: Q1 (January-March), Q2 (April-June), Q3 (July-September), and Q4 (October-December).

---

[2] The project area in this example is not realistic and has been simplified for demonstration purposes. Analyses of transit facilities will likely include inbound and outbound links beyond what is described in this simplified example, as well as the surrounding area.

[3] This decision is made to save time and effort, as it would result in the need for fewer modeling runs. More accurate results would be obtained by treating weekends differently and modeling them using the actual estimated Saturday and Sunday traffic.

Exhibit F-2. Average Weekday Bus Trips through Transit Terminal for Each Hour

Hour	Bus Trips
12am - 1am	7
1am - 2am	6
2am - 3am	6
3am - 4am	6
4am - 5am	7
5am - 6am	9
6am - 7am	27
7am - 8am	48
8am - 9am	39
9am - 10am	29
10am - 11am	26
11am - 12pm	28
12pm - 1pm	30
1pm - 2pm	31
2pm - 3pm	31
3pm - 4pm	39
4pm - 5pm	44
5pm - 6pm	42
6pm - 7pm	26
7pm - 8pm	21
8pm - 9pm	22
9pm - 10pm	17
10pm - 11pm	13
11pm - 12am	10

F.5.2   Deciding on how to handle link activity

As discussed in Section 4.2 of the guidance, MOVES offers several options for users to apply activity information to each LinkID. For illustrative purposes, based on the available information for the project (average speed, hourly bus volume, idle time, and fraction of vehicles encountering a red light) several methods of deriving Op-Mode distributions are employed in this example, as described below.

- Links 1 and 2 represent buses driving at an average of 5 mph through the terminal, entering and exiting the bus bays. An average speed of 5 mph is entered into the MOVES "links" input, which calculates an Op-Mode distribution to reflect the MOVES default 5 mph driving pattern.
- The queue link (Link 3) is given an Op-Mode distribution that represents buses decelerating, idling, and accelerating (red light) as well as cruising through (green light). First, an Op-Mode distribution is calculated for the link average speed (5 mph). Because this does not adequately account for idling at the intersection, the Op-Mode fractions are re-allocated to add in 50% idling (determined after consulting the *Highway Capacity Manual 2000* to approximate idle time in an under-capacity scenario) reflecting 50% of buses encountering a red light. A

fraction of 0.5 for Op-Mode "1" is added to the re-allocated 5 mph average speed Op-Mode distribution. The resulting Op-Mode distribution represents all activity on a queuing intersection link.

- The bus bays (Links 4 through 9) are represented by a single link (modeled in MOVES as "LinkID 4") and activity is defined in the Links table by an average speed of "0", representing exclusively idle activity.

F.5.3   Determining the number of MOVES runs

Following the guidance given in Section 4.3, it is determined that 16 MOVES runs should be completed to produce emission factors that show variation across four hourly periods (12 a.m., 6 a.m., 12 p.m., and 6 p.m., corresponding to overnight, morning, midday, and evening traffic scenarios, respectively) and four quarterly periods (represented by the months of January, April, July, and October; see Section 3.3). MOVES would calculate values for all project links for the time period specified in each run. Although traffic data is available for 24 hours, the emission factors produced from the 16 scenarios would be post-processed into grams/vehicle-hour and further converted to grams/hour emission factors that vary based on the hour-specific vehicle count. This methodology avoids running 24 hourly scenarios for four quarters (96 runs). Grams/hour emissions rates are required to use AERMOD.

F.5.4   Developing basic run specification inputs

When configuring MOVES for the analysis, the project sponsor follows Section 4.4 of the guidance, including, but not limited to, the following:

- From the Scale menu, selecting the "Project" domain; in addition, choosing output in "Inventory" so that total emission results are produced for each link, which is equivalent to a grams/hour/link emission factor needed by AERMOD (see Section 4.4.2).
- From the Time Spans panel, the appropriate year, month, day, and hour for each run is selected (see Section 4.4.3).
- From the Geographic Bounds panel, the custom domain is selected (see Section 4.4.4).
- From the Vehicles/Equipment panel, Diesel Transit Buses are selected (see Section 4.4.5).
- From the Road Types panel, the Urban Unrestricted road type is selected (see Section 4.4.6).
- From the Pollutants and Processes panel, appropriate pollutant/processes are selected according to Section 4.4.7 of the guidance for "highway links."
- In the Output panel, an output database is specified with grams and miles selected as units (see Section 4.4.10).

F.5.5   Entering project details using the Project Data Manager

<u>Meteorology</u>

As described previously, it is determined that MOVES should be run 16 times to reflect the following scenarios: 12 a.m., 6 a.m., 12 p.m., and 6 p.m. (corresponding to overnight, morning, midday, and evening traffic scenarios, respectively) for the months of January, April, July, and October. Temperature and humidity data from a representative meteorological monitoring station are obtained and confirmed to be consistent with data used in the regional emissions analysis from the currently conforming transportation plan and TIP (see Section 4.5.1). Average values for each hour and month combination are used for each of the 16 MOVES runs. As an example, temperature and humidity values for 12 a.m. January are shown in Exhibit F-3.

Exhibit F-3. Temperature and Humidity Input (January 6 a.m.)

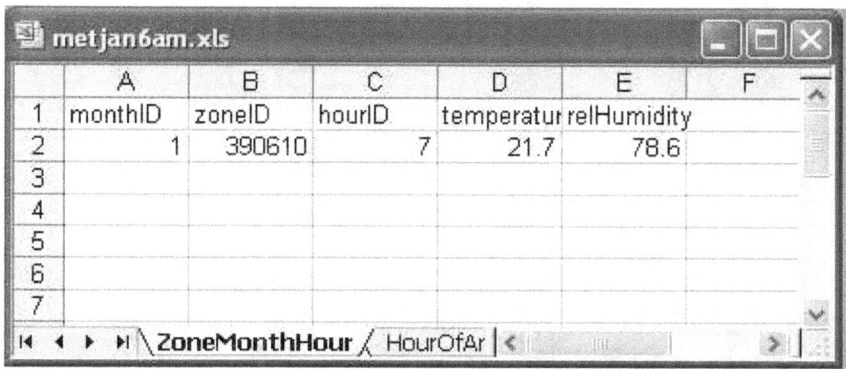

	A	B	C	D	E	F
1	monthID	zoneID	hourID	temperatur	relHumidity	
2	1	390610	7	21.7	78.6	
3						
4						
5						
6						
7						

ZoneMonthHour / HourOfAr

Age Distribution

Section 4.5.2 of the guidance specifies that default data should be used only if an alternative local dataset cannot be obtained and the regional conformity analysis relies on national defaults. However, for the sake of simplicity only, in this example the national default age distribution for 2015 is used for all vehicles and all runs (see Exhibit F-4). As discussed in the guidance, transit agencies should be able to provide a fleet-specific age distribution, and the use of fleet-specific data is always recommended (and would be expected in a real-world scenario) because emission factors vary significantly depending on the age of the fleet.

Exhibit F-4. Age Distribution Table (Partial)

	A	B	C	D	E	F
1	sourceTyp	yearID	ageID	ageFraction		
2	42	2015	0	0.052013		
3	42	2015	1	0.052432		
4	42	2015	2	0.051104		
5	42	2015	3	0.050951		
6	42	2015	4	0.0509		
7	42	2015	5	0.050341		
8	42	2015	6	0.045595		
9	42	2015	7	0.038191		
10	42	2015	8	0.034719		
11	42	2015	9	0.038183		
12	42	2015	10	0.043573		
13	42	2015	11	0.043438		
14	42	2015	12	0.051516		
15	42	2015	13	0.047071		
16	42	2015	14	0.043819		
17	42	2015	15	0.035929		
18	42	2015	16	0.031348		

agedist.xls — sourceTypeAgeDistribution

Fuel Supply and Fuel Formulation

An appropriate fuel supply and formulation is selected to match the project area's diesel use. In MOVES, diesel fuel formulation is constant across all quarters, so one fuel supply/fuel formulation combination is used for all MOVES runs. Also, it is known that all transit buses would use the same diesel fuel, so a fraction of 1 is entered for fuel 20011 (ultra-low-sulfur diesel fuel) in the Fuel Supply Table. In the case of this example, the default fuel supply/formulation matches the actual fuel supply/formulation, so it is therefore appropriate to use the default in the analysis (see Exhibits F-5 and F-6).

Exhibit F-5. Fuel Supply Table

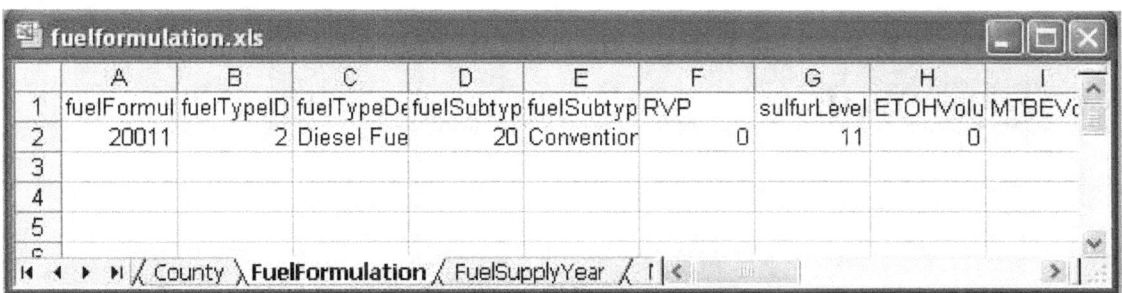

Exhibit F-6. Fuel Formulation Table

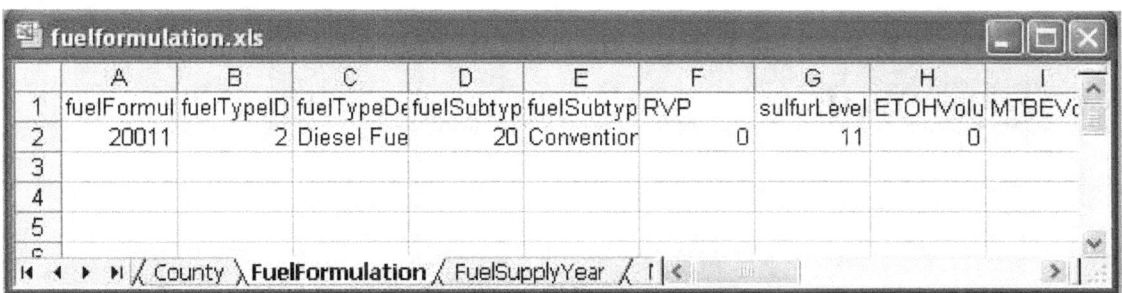

Inspection and Maintenance (I/M)

As there is no PM emissions benefit in MOVES for I/M programs, this menu item is skipped (see Section 4.5.4).

Link Source Type

The distribution of vehicle types on each link is defined in the Link Source Type table following the guidance in Section 4.5.5. Given that the project will be a dedicated transit bus terminal this analysis assumes only transit buses are operating on all links. Therefore, a fraction of 1 is entered for Source Type 42 (Transit Buses) for each LinkID indicating 100% of vehicles using the project are transit buses (see Exhibit F-7).

Exhibit F-7. Link Source Type Table

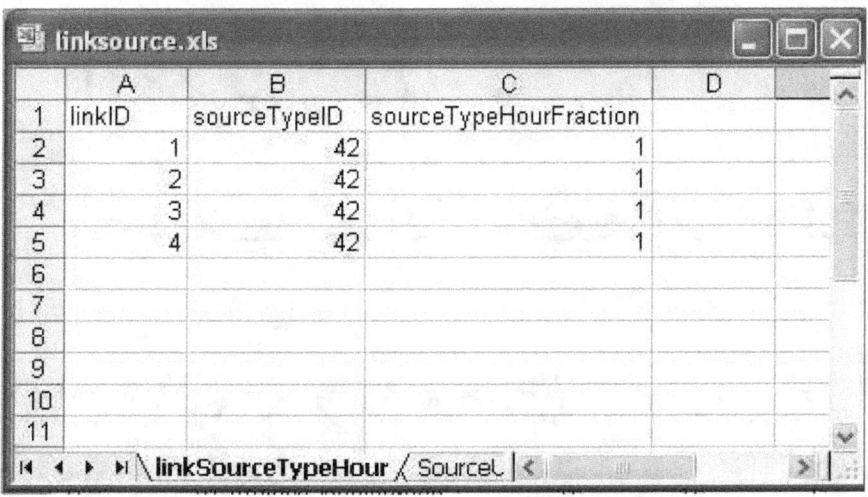

Links

The Links table (see Exhibit F-8) is populated with parameters for the four defined links of the bus terminal: three running links (Links 1-3) and one idle link (representing the terminal links). The link length is entered in terms of miles for each link. The road type for the four links is classified as "5" (Urban Unrestricted). The entrance and exit links (Links 1 and 2) are given an average speed of 5 mph. The queue link (Link 3) is given an average speed of 2.5 mph, representing 50% of the vehicle operating hours in idling mode and 50% operating hours traveling at 5 mph. Although MOVES is capable of calculating emissions from an average speed (as is done for Links 1 and 2), the specific activity on Link 3 is directly entered with an Op-Mode distribution. LinkID 4 is given a link average speed of "0" mph, which indicates entirely idle operation. Link volume (which represents the number of buses per hour) is entered for each link; however, since the goal of the analysis is to produce an estimate in grams/vehicle-hour, the volume (i.e., the number of vehicles) will be divided out during post-processing. Also, because link volume is arbitrary, the Links table shown in Exhibit F-8 can be used for all 16 MOVES runs.

Exhibit F-8. Links Table

	A	B	C	D	E	F	G	H
1	linkID	countyID	zoneID	roadTypeID	linkLength	linkVolume	linkAvgSpeed	linkDescription
2	1	99001	990010	5	0.038	48	5	Entrance Link
3	2	99001	990010	5	0.030	24	5	Exit Approach Link
4	3	99001	990010	5	0.008	24	2.5	Left Turn Exit Link
5	4	99001	990010	5	0.011	8	0	Terminals

Describing Vehicle Activity

MOVES can capture details about vehicle activity in a number of ways. In this case, it is decided to use average speeds for Links 1, 2, and 4 and a detailed Op-Mode distribution for Link 3 (see Section 4.5.7).

Op-Mode distributions for Links 1 and 2 are calculated within MOVES based on a 5 mph average speed. The MOVES model calculates a default Op-Mode distribution based on average speed and road type (for these links, 5 mph on road type 5). Link 3 is given a unique Op-Mode distribution to better simulate the queuing and idling that occurs prior to buses exiting the facility at a traffic signal. The sponsor estimates that 50% of buses would idle at a red light before exiting the facility, so the idling operation (OpMode ID 1) is assumed to be 0.5 for Link 3. The remaining 50% is re-allocated based on the default 5 mph Op-Mode distribution calculated for Links 1 and 2 (which includes acceleration, deceleration, and cruise operating modes). This process requires an additional MOVES run to extract the default 5 mph Op-Mode distribution from the MOVES execution database. By selecting "save data" for the "Operating Mode Distribution Generator (Running OMDG)" under the MOVES "Advanced Performance Features" panel, the Op-Mode distributions generated for 5 mph on an urban unrestricted road type are saved in the MOVES execution database in the MySQL table "opmodedistribution." The Op-Mode distribution used in the analysis for Link 3 is partially shown in Exhibit F-9.

Exhibit F-9. Link 3 (Queue Link) Op-Mode Distribution Input Table (Partial)

	A	B	C	D	E	F	G	H	I
1	sourceTyp	hourDayID	linkID	polProces	opModelD	opModeFraction			
2	42	75	3	9101	1	0.5			
3	42	75	3	9190	1	0.5			
4	42	75	3	11001	1	0.5			
5	42	75	3	11015	1	0.5			
6	42	75	3	11017	1	0.5			
7	42	75	3	11090	1	0.5			
8	42	75	3	11101	1	0.5			
9	42	75	3	11115	1	0.5			
10	42	75	3	11117	1	0.5			
11	42	75	3	11190	1	0.5			
12	42	75	3	11201	1	0.5			
13	42	75	3	11215	1	0.5			
14	42	75	3	11217	1	0.5			
15	42	75	3	11290	1	0.5			
16	42	75	3	11501	1	0.5			
17	42	75	3	11515	1	0.5			
18	42	75	3	11517	1	0.5			
19	42	75	3	11590	1	0.5			
20	42	75	3	11609	1	0.5			
21	42	75	3	11710	1	0.5			
22	42	75	3	9101	11	0.25			
23	42	75	3	9190	11	0.25			
24	42	75	3	11001	11	0.25			
25	42	75	3	11015	11	0.25			
26	42	75	3	11017	11	0.25			
27	42	75	3	11090	11	0.25			
28	42	75	3	11101	11	0.25			
29	42	75	3	11115	11	0.25			
30	42	75	3	11117	11	0.25			
31	42	75	3	11190	11	0.25			
32	42	75	3	11201	11	0.25			

opModeDistribution / HourDay / OperatingMode

Off-Network

As it is assumed that there are no off-network links (such as parking lots or truck stops) that would have to be considered using the Off-Network Importer (bus idling at the terminal is captured by the terminal links), there is no need to use this option in this example. As noted earlier, this assumption is made to simplify the example. Most transit projects would include rider parking lots and should include these emissions in a PM hot-spot analysis.

F.5.6 Generating emission factors for use in air quality modeling

After generating the run specification and entering the required information into the Project Data Manager as described above, MOVES is run 17 times: 16 runs (four hours of the day for four quarters of the year) plus an initial run to generate the Op-Mode distribution for 5 mph as discussed earlier. Upon completion of each run, the MOVES output is located in the MySQL output database table "movesoutput" and sorted by Month, Hour, LinkID, ProcessID, and PollutantID. An aggregate $PM_{2.5}$ emission factor is then calculated by the project sponsor for each Month, Hour, and LinkID combination using the following equation and the guidance given in Section 4.4.7 of the guidance:[4]

$$PM_{\text{aggregate total}} = (PM_{\text{total running}}) + (PM_{\text{total crankcase running}}) + (\text{brake wear}) + (\text{tire wear})$$

For each link, the total emissions are divided by the number of vehicles on each link (as reported in the "movesactivityoutput" table ActivitytypeID = 6) to produce a grams/vehicle-hour value. This value is then multiplied by the number of buses on each link, for each of the 24 hours where data are available (see Exhibit F-2).

The emission factor (grams/vehicle-hour) for LinkID 4 (links 4 through 9) is converted into grams per vehicle-minute, and then multiplied by the total idle time for each unique hour. For instance, the hour from 5 p.m. to 6 p.m. has a volume of 42 buses per hour (7 buses per bus bay). If each bus is expected to idle for 60 seconds each hour, the total idle time for each bus bay for that hour would be 7 minutes per hour. If MOVES calculated a PM emission factor of 2.0 grams per vehicle-minute, the emission factor for each bus bay link under this scenario would be 14.0 grams/hour.

To account for temperature changes throughout the day, emission factors are evenly paired with corresponding traffic volumes (six hours per period):
- 6 a.m. results – traffic data from 3 a.m. to 9 a.m.
- 12 p.m. results – traffic data from 9 a.m. to 3 p.m.
- 6 p.m. results – traffic data from 3 p.m. to 9 p.m.
- 12 p.m. results – traffic data from 9 p.m. to 3 a.m.

The emission factor results for each quarter are similarly paired with traffic volumes.

---

[4] EPA is considering creating one or more MOVES scripts that would automate the summing of aggregate emissions when completing project-level analyses. These scripts would be made available for download on the MOVES website (www.epa.gov/otaq/models/moves/tools.htm), when available.

The 96 resulting grams/hour emission factors (24 hours each for four quarters) for each link are then ready to be used as an input to the AERMOD dispersion model to predict $PM_{2.5}$ concentrations.

F.6   ESTIMATE EMISSIONS FROM ROAD DUST, CONSTRUCTION AND OTHER ADDITIONAL SOURCES (STEP 4)

F.6.1   Estimating re-entrained road dust

In this case, this area does not have any adequate or approved SIP budgets for either $PM_{2.5}$ NAAQS, and neither the EPA nor the state air agency has made a finding that road dust emissions are a significant contributor to the air quality problem for either $PM_{2.5}$ NAAQS. Therefore, $PM_{2.5}$ emissions from road dust do not need to be considered in this analysis (see Sections 2.5.3 and 6.2).

F.6.2   Estimating transportation-related construction dust

The construction of this project will not occur during the analysis year. Therefore, emissions from construction dust are not included in this analysis (see Sections 2.5.5 and 6.4).

F.6.3   Estimating additional sources of emissions in the project area

Through interagency consultation, it is determined that the project area in the analysis year does not include locomotives or other nearby emissions sources that would have to be included in the air quality modeling (see Section 6.6).

F.7   SELECT AN AIR QUALITY MODEL, DATA INPUTS, AND RECEPTORS (STEP 5)

F.7.1   Characterizing emission sources

Because this is a transit terminal project, EPA's AERMOD model is determined to be the appropriate dispersion model to use for this analysis (see Section 7.3). AERMOD is run to estimate $PM_{2.5}$ concentrations in and around the bus terminal project. Each link is represented in AERMOD as an "Area Source" with dimensions matching the project description (see Exhibit F-1). The emission release height is set to three meters, the approximate exhaust height of most transit buses.

Emission factors generated from the MOVES runs are added to the AERMOD input file (see Exhibit F-10). For this analysis, emissions vary significantly from hour to hour due to fluctuating bus volumes as well as from daily and quarterly temperature effects. Adjustment factors (EMISFACT) are used to model these hourly and quarterly variations in emission factors. Refer to Appendix J for additional detail on using hour-by-hour emission differences in an AERMOD input file.

Exhibit F-10. AERMOD Input File (Partial) with Seasonal (Quarterly) and Hourly Adjustment Factors (Circled)

AERMOD Input.txt - Notepad

File Edit Format View Help

```
CO STARTING
CO TITLEONE  Transit Example
CO MODELOPT  FLAT  CONC
CO RUNORNOT  RUN
CO AVERTIME  24  ANNUAL
CO POLLUTID  OTHER
CC FINISHED

SC STARTING
SO ELEVUNIT  METERS
SO LOCATION  LINK1      AREA       -166.7  -50.9  0
SO LOCATION  LINK2      AREA       -166.6  -33.2  0
SO LOCATION  LINK3      AREA       -118.1  -33.2  0
SO LOCATION  LINK5      AREAPOLY   -156.1  -47.4  0
SO LOCATION  LINK4      AREAPOLY   -163.5  -47.4  0
SO LOCATION  LINK7      AREAPOLY   -140.5  -47.4  0
SO LOCATION  LINK8      AREAPOLY   -132.9  -47.4  0
SO LOCATION  LINK9      AREAPOLY   -125.3  -47.3  0
SO LOCATION  LINK6      AREAPOLY   -148   -47.4  0
SO SRCPARAM  LINK1      7E-06  3  60.6  3.6  0.0  0
SO SRCPARAM  LINK2      4E-06  3  48.5  3.6  0.0  0
SO SRCPARAM  LINK3      4E-06  3  12.1  3.6  0.0  0
SO SRCPARAM  LINK5      2E-06  3  4  0
SO SRCPARAM  LINK4      2E-06  3  4  0
SO SRCPARAM  LINK7      2E-06  3  4  0
SO SRCPARAM  LINK8      2E-06  3  4  0
SO SRCPARAM  LINK9      2E-06  3  4  0
SO SRCPARAM  LINK6      2E-06  3  4  0
SO AREAVERT  LINK5      -156.1 -47.4  -152.5 -47.4  -141.3 -33.3  -144.7 -33.3
SO AREAVERT  LINK4      -163.5 -47.4  -159.9 -47.4  -148.7 -33.3  -152.1 -33.3
SO AREAVERT  LINK7      -140.5 -47.4  -136.9 -47.4  -125.7 -33.3  -129.1 -33.3
SO AREAVERT  LINK8      -132.9 -47.4  -129.3 -47.4  -118.1 -33.3  -121.5 -33.3
SO AREAVERT  LINK9      -125.3 -47.3  -121.7 -47.3  -110.5 -33.2  -113.9 -33.2
SO AREAVERT  LINK6      -148.0 -47.4  -144.4 -47.4  -133.2 -33.3  -136.6 -33.3
SO EMISFACT  LINK1      SEASHR  0.37  0.44  0.51  0.58  0.65  0.72  0.79  0.86
SO EMISFACT  LINK1      SEASHR  0.93  1  0.93  0.86  0.79  0.72  0.65  0.72  0.79
SO EMISFACT  LINK1      SEASHR  0.86  1  0.93  0.86  0.79  0.72  0.65  0.37  0.44
SO EMISFACT  LINK1      SEASHR  0.51  0.58  0.65  0.72  0.79  0.86  0.93  1  0.93
SO EMISFACT  LINK1      SEASHR  0.86  0.79  0.72  0.65  0.72  0.79  0.86  1  0.93
SO EMISFACT  LINK1      SEASHR  0.86  0.79  0.72  0.65  0.37  0.44  0.51  0.58
SO EMISFACT  LINK1      SEASHR  0.65  0.72  0.79  0.86  0.93  1  0.93  0.86  0.79
SO EMISFACT  LINK1      SEASHR  0.72  0.65  0.72  0.79  0.86  1  0.93  0.86  0.79
SO EMISFACT  LINK1      SEASHR  0.72  0.65  0.37  0.44  0.51  0.58  0.65  0.72
SO EMISFACT  LINK1      SEASHR  0.79  0.86  0.93  1  0.93  0.86  0.79  0.72  0.65
SO EMISFACT  LINK1      SEASHR  0.72  0.79  0.86  1  0.93  0.86  0.79  0.72  0.65
SO EMISFACT  LINK2      SEASHR  0.37  0.44  0.51  0.58  0.65  0.72  0.79  0.86
SO EMISFACT  LINK2      SEASHR  0.93  1  0.93  0.86  0.79  0.72  0.65  0.72  0.79
SO EMISFACT  LINK2      SEASHR  0.86  1  0.93  0.86  0.79  0.72  0.65  0.37  0.44
SO EMISFACT  LINK2      SEASHR  0.51  0.58  0.65  0.72  0.79  0.86  0.93  1  0.93
SO EMISFACT  LINK2      SEASHR  0.86  0.79  0.72  0.65  0.72  0.79  0.86  1  0.93
SO EMISFACT  LINK2      SEASHR  0.86  0.79  0.72  0.65  0.37  0.44  0.51  0.58
SO EMISFACT  LINK2      SEASHR  0.65  0.72  0.79  0.86  0.93  1  0.93  0.86  0.79
SO EMISFACT  LINK2      SEASHR  0.72  0.65  0.72  0.79  0.86  1  0.93  0.86  0.79
SO EMISFACT  LINK2      SEASHR  0.72  0.65  0.37  0.44  0.51  0.58  0.65  0.72
SO EMISFACT  LINK2      SEASHR  0.79  0.86  0.93  1  0.93  0.86  0.79  0.72  0.65
```

F.7.2 Incorporating meteorological data

A representative set of meteorology data, as well as an appropriate surface roughness are selected (see Section 7.5). The recommended five years of meteorological data is obtained from a local airport for calendar years 1998-2002. The appropriate surface roughness is set at 1 meter, consistent with the recommendations made in the AERMOD Implementation Guide.

F.7.3 Placing receptors

Using the guidance given in Section 7.6, receptors are placed at appropriate locations within the area substantially affected by the project (see Exhibit F-11).[5] It is determined in this instance to locate receptors around the perimeter of the project in increments of five meters as well as within the passenger loading areas adjacent to the bus bays. Receptor heights are set at 1.8 meters. A background concentration of "0" is input into the model. Representative background concentrations are added at a later step (see Step 7).

AERMOD is run using five years of meteorological data and output produced for all receptors for each of the five years of meteorological data.

Exhibit F-11. Area Source and Receptor Locations for Air Quality Modeling

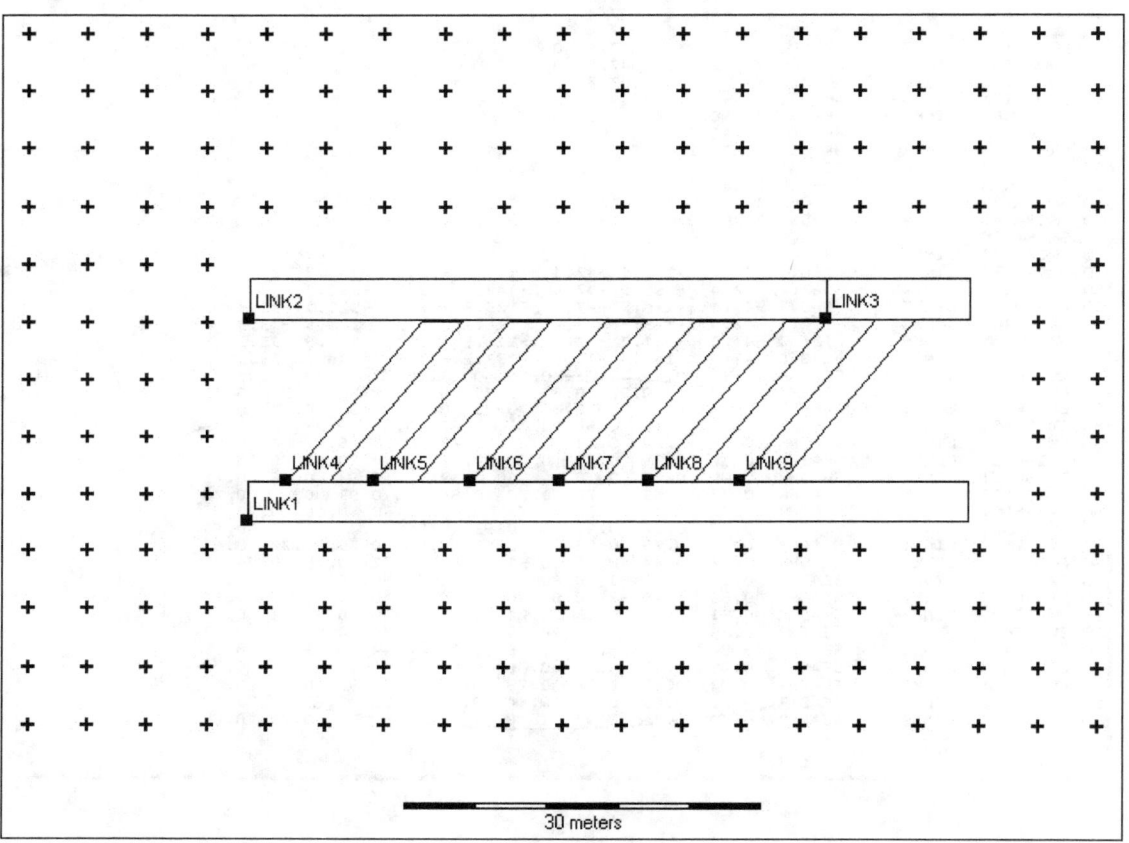

---

[5] The number and arrangement of receptors used in this example are simplified for ease of explanation; real-world projects could expect to see more receptors and include the surrounding area.

F.8 DETERMINE BACKGROUND CONCENTRATIONS FROM NEARBY AND OTHER EMISSION SOURCES (STEP 6)

Through the interagency consultation process, a nearby upwind $PM_{2.5}$ monitor that has been collecting ambient data for both the annual and 24-hour $PM_{2.5}$ NAAQS is determined to be representative of the background air quality at the project location (see Exhibit F-12). The most recent data set is used (in this case, calendar year 2008 through 2010) and average 24-hour $PM_{2.5}$ values are provided in a four-day/three-day measurement interval. As previously noted, no nearby sources requiring modeling are identified.

Note: This is a highly simplified situation for illustrative purposes; refer to Section 8 of the guidance for additional considerations for how to most accurately reflect background concentrations in a real-world scenario.

Exhibit F-12. $PM_{2.5}$ Monitor Data from a Representative Nearby Site (Partial)

	A	B	C	D	E	F	G
1	Month	Day	Year	Date	PM2.5 Concentration		
2	1	1	2008	1/1/2008	23.08		
3	1	5	2008	1/5/2008	5.69		
4	1	8	2008	1/8/2008	12.19		
5	1	12	2008	1/12/2008	6.71		
6	1	15	2008	1/15/2008	7.26		
7	1	19	2008	1/19/2008	17.92		
8	1	22	2008	1/22/2008	11.90		
9	1	26	2008	1/26/2008	14.37		
10	1	29	2008	1/29/2008	16.54		
11	2	2	2008	2/2/2008	7.40		
12	2	5	2008	2/5/2008	13.63		
13	2	9	2008	2/9/2008	19.15		
14	2	12	2008	2/12/2008	12.65		
15	2	16	2008	2/16/2008	14.77		
16	2	19	2008	2/19/2008	11.08		
17	2	23	2008	2/23/2008	18.00		
18	2	26	2008	2/26/2008	21.62		
19	3	1	2008	3/1/2008	14.65		
20	3	4	2008	3/4/2008	6.93		
21	3	8	2008	3/8/2008	19.03		
22	3	11	2008	3/11/2008	20.66		
23	3	15	2008	3/15/2008	11.99		
24	3	18	2008	3/18/2008	4.71		
25	3	22	2008	3/22/2008	11.05		
26	3	25	2008	3/25/2008	15.64		
27	3	29	2008	3/29/2008	6.64		
28	4	1	2008	4/1/2008	11.68		
29	4	5	2008	4/5/2008	5.04		
30	4	8	2008	4/8/2008	10.11		
31	4	12	2008	4/12/2008	11.96		

backgroundcalcs.xls

Monitoring Data / AERMOD 24-hour Output

F.9   CALCULATE DESIGN VALUES AND DETERMINE CONFORMITY (STEP 7)

With both MOVES outputs and background concentrations now available, the project sponsor can calculate the design values. For illustrative purposes, calculations for a single receptor with the highest modeled concentrations for the build scenario are shown in this example. [6] In Step 7, the guidance from Sections 9.3.2 and 9.3.3 are used to calculate design values from the modeled results and the background concentrations for comparison with the annual and 24-hour $PM_{2.5}$ NAAQS.

F.9.1   Determining conformity to the annual $PM_{2.5}$ NAAQS

First, average background concentrations are determined for each year of monitored data (shown in Exhibit F-13). The three-year average background concentration is then calculated (see Exhibit F-14).

Exhibit F-13. Annual Average Background Concentration for Each Year

Monitoring Year	Annual Average Background Concentration
2008	13.348
2009	12.785
2010	13.927
Annual Average	13.353

Exhibit F-14. Calculation of Annual Design Value (At Highest Receptor)

Annual Average Background Concentration (Three-year Average)	Annual Average Modeled Concentration (Five-year Average)	Sum of Background + Project	Annual Design Value
13.353	1.423	14.776	14.8

---

[6] In an actual PM hot-spot analysis, design values would be calculated at additional receptors as described in Section 9.3.

To determine the annual $PM_{2.5}$ design value, the annual average background concentration is added to the five-year annual average modeled concentration (at the receptor with the highest annual average concentration from the AERMOD output). This calculation is shown in Exhibit F-14. The sum (project + background) results in a design value of 14.8 $\mu g/m^3$. This value at the highest receptor is less than the 1997 annual $PM_{2.5}$ NAAQS of 15.0 $\mu g/m^3$. It can be assumed that all other receptors with lower modeled concentrations will also have design values less than the 1997 annual $PM_{2.5}$ NAAQS. In this example it is unnecessary to determine appropriate receptors in the build scenario or develop a no-build scenario for the annual $PM_{2.5}$ NAAQS, since the build scenario demonstrates that the hot-spot analysis requirements in the transportation conformity rule are met at all receptors.

F.9.2   *Determining conformity to the 24-Hour $PM_{2.5}$ NAAQS*

The next step is to calculate a design value to compare with the 2006 24-hour $PM_{2.5}$ NAAQS through a "Second Tier" analysis as described in Section 9.3.3. For ease of explanation, this process has been divided into individual steps, consistent with the guidance.

Step 7.1
The number of background measurements is counted for each year of monitored data (2008 to 2010). Based on a 4-day/3-day measurement interval, the dataset has 104 values per year.

Step 7.2
For each year of monitored concentrations, the eight highest daily background concentrations for each quarter are determined, resulting in 32 values (4 quarters; 8 concentrations/quarter) for each year of data (shown in Exhibit F-15).

Step 7.3
Identify the highest modeled 24-hour concentration in each quarter, averaged across each year of meteorological data. For illustrative purposes, the highest average concentration across five years of meteorological data for a single receptor in each quarter is shown in Exhibit F-16. Note that, in a real-world situation, this process would be repeated for all receptors in the build scenario.

Exhibit F-15. Highest Daily Background Concentrations for Each Quarter and Each Year

2008				
Rank	Q1	Q2	Q3	Q4
1	20.574	21.262	22.354	20.434
2	20.152	20.823	22.042	20.016
3	19.743	20.398	21.735	19.611
4	19.346	19.985	21.434	19.218
5	18.961	19.584	21.140	18.837
6	18.588	19.196	20.851	18.467
7	18.226	18.819	20.568	18.109
8	17.874	18.454	20.291	17.761
2009				
Rank	Q1	Q2	Q3	Q4
1	20.195	20.867	21.932	20.058
2	19.784	20.440	21.628	19.651
3	19.386	20.026	21.329	19.257
4	19.000	19.624	21.037	18.875
5	18.625	19.235	20.750	18.504
6	18.262	18.857	20.469	18.145
7	17.910	18.490	20.194	17.796
8	17.568	18.135	19.924	17.457
2010				
Rank	Q1	Q2	Q3	Q4
1	21.137	21.847	22.980	20.990
2	20.698	21.390	22.655	20.556
3	20.272	20.948	22.336	20.135
4	19.860	20.519	22.023	19.726
5	19.459	20.102	21.717	19.330
6	19.071	19.698	21.417	18.945
7	18.694	19.307	21.123	18.572
8	18.329	18.927	20.834	18.211

Exhibit F-16. Five-year Average of Highest Modeled Concentrations for Each Quarter (At Example Receptor)

	Q1	Q2	Q3	Q4
Five Year Average Maximum Concentration (At Example Receptor)	6.51	6.64	6.71	6.63

Step 7.4

The highest modeled concentration in each quarter (from Step 7.3) is added to each of the eight highest monitored concentrations for the same quarter for each year of monitoring data (from Step 7.2). As shown in Exhibit F-17, this step results in eight concentrations in each of four quarters for a total of 32 values for each year of monitoring data. As mentioned, this example analysis shows only a single receptor's values, but project sponsors should calculate design values at all receptors in the build scenario.

Exhibit F-17. Sum of Background and Modeled Concentrations at Example Receptor for Each Quarter

2008				
	Q1	Q2	Q3	Q4
1	27.088	27.901	29.063	26.948
2	26.667	27.462	28.750	26.530
3	26.258	27.037	28.443	26.125
4	25.861	26.624	28.143	25.732
5	25.476	26.224	27.848	25.351
6	25.102	25.835	27.560	24.982
7	24.740	25.459	27.277	24.623
8	24.389	25.093	27.000	24.275
2009				
	Q1	Q2	Q3	Q4
1	26.709	27.506	28.641	26.572
2	26.298	27.079	28.336	26.166
3	25.900	26.665	28.038	25.772
4	25.514	26.264	27.745	25.389
5	25.140	25.874	27.459	25.019
6	24.776	25.496	27.178	24.659
7	24.424	25.130	26.903	24.310
8	24.082	24.774	26.633	23.971
2010				
	Q1	Q2	Q3	Q4
1	27.651	28.486	29.689	27.505
2	27.212	28.030	29.363	27.070
3	26.787	27.587	29.044	26.649
4	26.374	27.158	28.732	26.240
5	25.974	26.742	28.426	25.844
6	25.585	26.338	28.125	25.460
7	25.209	25.946	27.831	25.087
8	24.843	25.566	27.543	24.725

Step 7.5

As shown in Exhibit F-18, for each year of monitoring data, the 32 values from Step 7.4 are ordered together in a column and assigned a yearly rank for each value, from 1 (highest concentration) to 32 (lowest concentration).

Exhibit F-18. Ranking Sum of Background and Modeled Concentrations at Example Receptor for Each Year of Background Data

Rank	2008	2009	2010
1	29.063	28.641	29.689
2	28.750	28.336	29.363
3	**28.443**	**28.038**	**29.044**
4	28.143	27.745	28.732
5	27.901	27.506	28.486
6	27.848	27.459	28.426
7	27.560	27.178	28.125
8	27.462	27.079	28.030
9	27.277	26.903	27.831
10	27.088	26.709	27.651
11	27.037	26.665	27.587
12	27.000	26.633	27.543
13	26.948	26.572	27.505
14	26.667	26.298	27.212
15	26.624	26.264	27.158
16	26.530	26.166	27.070
17	26.258	25.900	26.787
18	26.224	25.874	26.742
19	26.125	25.772	26.649
20	25.861	25.514	26.374
21	25.835	25.496	26.338
22	25.732	25.389	26.240
23	25.476	25.140	25.974
24	25.459	25.130	25.946
25	25.351	25.019	25.844
26	25.102	24.776	25.585
27	25.093	24.774	25.566
28	24.982	24.659	25.460
29	24.740	24.424	25.209
30	24.623	24.310	25.087
31	24.389	24.082	24.843
32	24.275	23.971	24.725

Step 7.6
For each year of monitoring data, the value with a rank that corresponds to the projected 98th percentile concentration is determined. As discussed in Section 9, an analysis employing 101-150 background values for each year (as noted in Step 7.1, this analysis uses 104 values per year) uses the 3rd highest rank to represent a 98th percentile. The 3rd highest concentration (highlighted in Exhibit F-18) is referred to as the "projected 98th percentile concentration."

Step 7.7
Steps 7.1 through 7.6 are repeated to calculate a projected 98th percentile concentration at each receptor based on each year of monitoring data and modeled concentrations.

Step 7.8
For the example receptor, the average of the three projected 98th percentile concentrations (highlighted in Exhibit F-18) is calculated.

Step 7.9
The resulting value of 28.508 $\mu g/m^3$ is then rounded to the nearest whole $\mu g/m^3$ resulting in a design value at the example receptor of 29 $\mu g/m^3$. At each receptor this process should be repeated. However, in the case of this analysis, the example receptor is the receptor with the highest design value in the build scenario.

Step 7.10
The design values calculated at each receptor are compared to the NAAQS. In the case of this example, the highest 24-hour design value (29 $\mu g/m^3$) is less than the 2006 24-hour $PM_{2.5}$ NAAQS of 35 $\mu g/m^3$. Since this is the design value at the highest receptor, it can be assumed that the conformity requirements are met at all receptors in the build scenario. Therefore, it is unnecessary for the project sponsor to calculate design values for the no-build scenario for the 24-hour $PM_{2.5}$ NAAQS.

F.10   CONSIDER MITIGATION AND CONTROL MEASURES (STEP 8)

In this case, the project is determined to conform. In situations when this is not the case, it may be necessary to consider additional mitigation or control measures. If measures are considered, additional air quality modeling would need to be completed and new design values calculated to ensure that conformity requirements are met. See Section 10 for more information, including some specific measures that might be considered.

F.11   DOCUMENT THE PM HOT-SPOT ANALYSIS (STEP 9)

The final step is to properly document the PM hot-spot analysis in the conformity determination (see Section 3.10).

Appendix G:
Example of Using EMFAC2011 for a Highway Project

G.1 INTRODUCTION

The purpose of this appendix is to demonstrate the procedures described in Section 5 of the guidance on using EMFAC2011 to generate emission factors for air quality modeling. The following example, based on a hypothetical, simplified highway project, illustrates the modeling steps required for users to run the EMFAC2011-PL tool to develop project-specific PM running exhaust emission factors using the "simplified approach" described in Section 5.5 of the guidance.

As discussed in the guidance, application of the simplified approach and use of the EMFAC2011-PL tool is only appropriate when the project-specific fleet age distribution does not differ from the EMFAC2011 defaults and the project does not include start or idling emissions. See Appendix H for an example of using the detailed approach to modify a default age distribution.

Users will be able to generate running emission factors (in grams/vehicle-mile) in a single EMFAC2011-PL run; multiple links and calendar years can also be handled within one run. This example does not include the subsequent air quality modeling; refer to Appendix E for an example of how to run an air quality model for a highway project for PM hot-spot analyses.

G.2 PROJECT CHARACTERISTICS

The hypothetical highway project is located in Sacramento County, California. For illustrative purposes, the project is characterized by a single link with an average link travel speed for all traffic equal to 65 mph.[1] Project-specific age distributions do not differ from the EMFAC2011 defaults, so a simplified modeling approach using the EMFAC2011-PL tool will be used to develop a link-specific $PM_{2.5}$ emission rate.

The project's first full year of operation is assumed to be the year 2013. Through the interagency consultation process, it is determined that 2015 should be the analysis year (based on the project's emissions and background concentrations). The build scenario 2015 traffic data for this highway project shows that 25% of the total project VMT is from trucks and 75% from non-trucks. This truck/non-truck fleet mix will be used to post-process the EMFAC-PL output.

---

[1] These are simplified data to illustrate the use of EMFAC2011; this example does not, for instance, separate data by peak vs. off-peak periods, divide the project into separate links, or consider additional analysis years, all of which would likely be required for an actual project.

G.3 DESCRIBING THE SCENARIO USING THE EMFAC2011-PL TOOL

Based on the project characteristics, it is first necessary to describe the modeling scenario in the EMFAC2011-PL interface (see Exhibits G-1 and G-2).

Exhibit G-1. Basic Inputs in EMFAC2011-PL for the Hypothetical Highway Project

Step	Input Category	Input Data	Note
1	Vehicle Category Scheme	Truck / Non-Truck Categories	Provides rates for truck/non-truck categories
2	Region type	County	Per Section 5.5.2 of the guidance
3	Region	Sacramento	Select from drop-down list
4	CalYr	2015	Select from drop-down list
5	Season	Annual	Select from drop-down list
6	Vehicle Category	ALL	Provides rates for HD and LD
7	Fuel Type	TOT	Does not generate separate rates for gasoline and diesel
8	Speed	65 MPH	Select from drop-down list

Exhibit G-2. EMFAC2011-PL GUI Showing Selections Made for the Hypothetical Highway Project

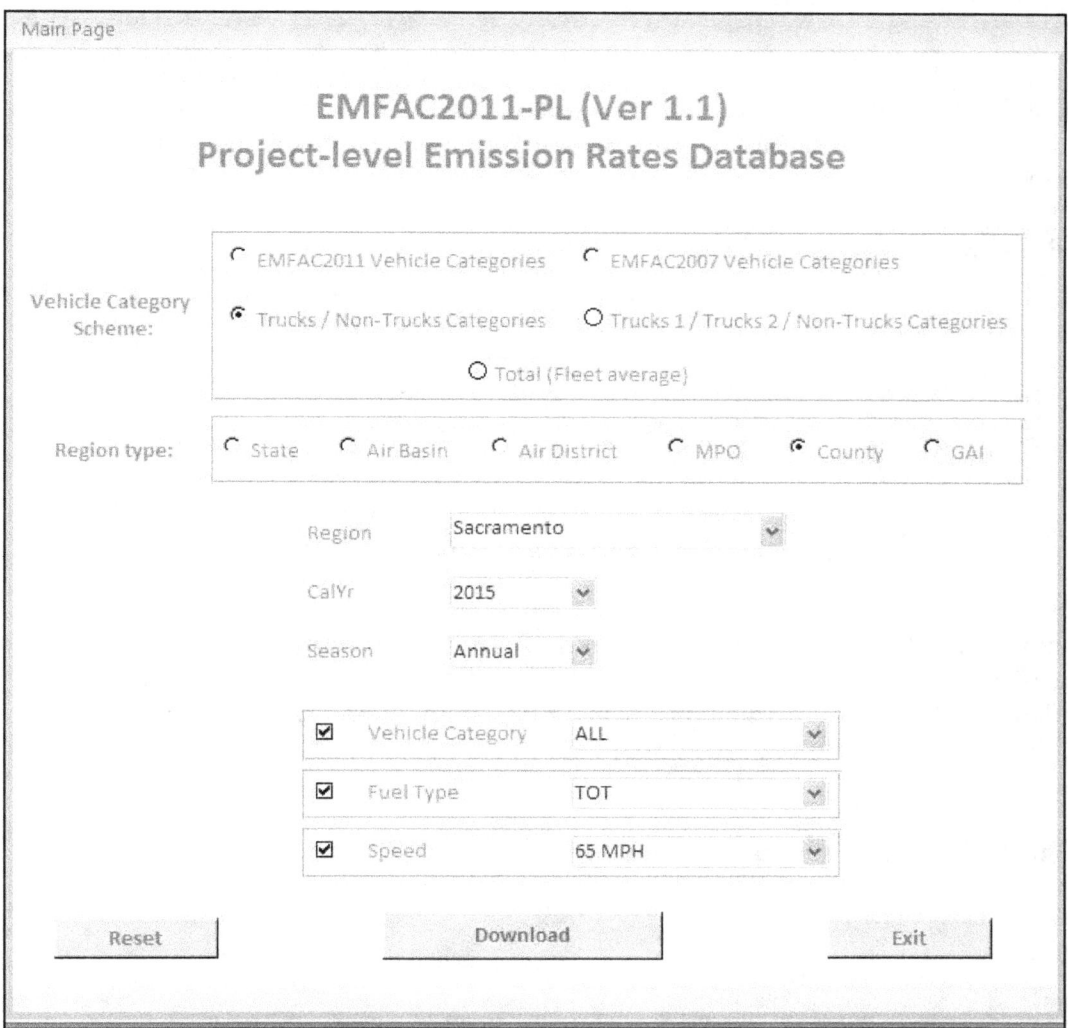

G.4 Calculating a Link-Specific Emission Rate from EMFAC2011-PL Output

After running EMFAC2011-PL, an output Excel file (Exhibit G-3) is produced in the EMFAC2011-PL folder. From this file, emission rates are appropriately processed to calculate a single link emission rate appropriate for dispersion modeling. This process is described below.

Exhibit G-3. EMFAC2011-PL Output File

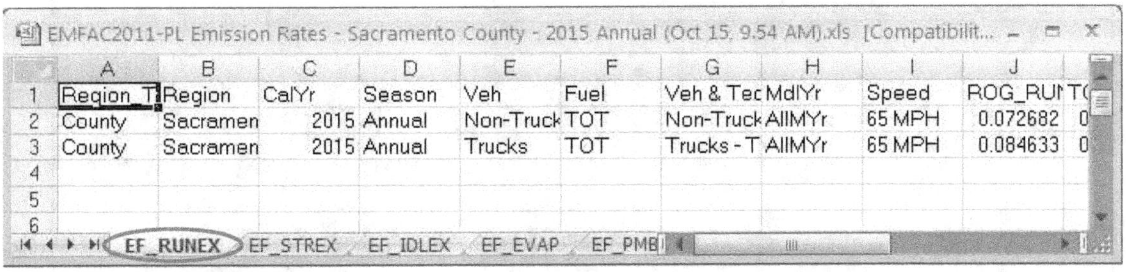
EMFAC2011-PL Emission Rates - Sacramento County - 2015 Annual (Oct 15, 9.54 AM).xls [Compatibilit...

	A	B	C	D	E	F	G	H	I	J
1	Region_T	Region	CalYr	Season	Veh	Fuel	Veh & Tec	MdlYr	Speed	ROG_RUI
2	County	Sacramen	2015	Annual	Non-Truck	TOT	Non-Truck	AllMYr	65 MPH	0.072682
3	County	Sacramen	2015	Annual	Trucks	TOT	Trucks - T	AllMYr	65 MPH	0.084633

EF_RUNEX | EF_STREX | EF_IDLEX | EF_EVAP | EF_PMB

The next step is to extract the relevant emission rates for post-processing in a separate Excel worksheet. For running emissions, the Total $PM_{2.5}$ emission factor (EF) is calculated as the sum of the running exhaust EF (Exhibit G-4), the brake wear EF, and the tire wear EF (Exhibit G-5).

Exhibit G-4. Running Exhaust Rates

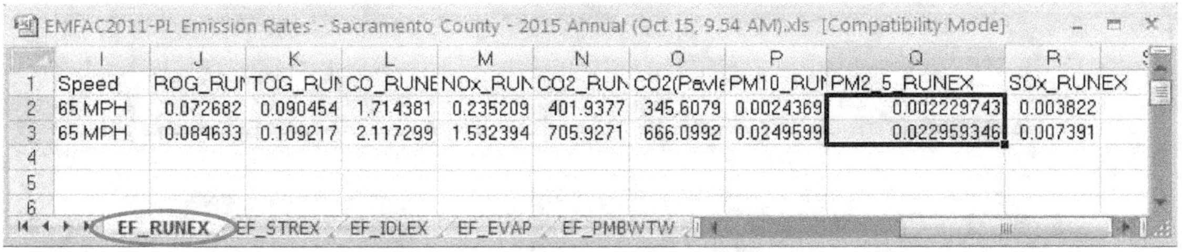
EMFAC2011-PL Emission Rates - Sacramento County - 2015 Annual (Oct 15, 9.54 AM).xls [Compatibility Mode]

	I	J	K	L	M	N	O	P	Q	R
1	Speed	ROG_RUI	TOG_RUI	CO_RUNE	NOx_RUN	CO2_RUN	CO2(Pavle	PM10_RUI	PM2_5_RUNEX	SOx_RUNEX
2	65 MPH	0.072682	0.090454	1.714381	0.235209	401.9377	345.6079	0.0024369	0.002229743	0.003822
3	65 MPH	0.084633	0.109217	2.117299	1.532394	705.9271	666.0992	0.0249599	0.022959346	0.007391

EF_RUNEX | EF_STREX | EF_IDLEX | EF_EVAP | EF_PMBWTW

Exhibit G-5. Brake Wear and Tire Wear Rates

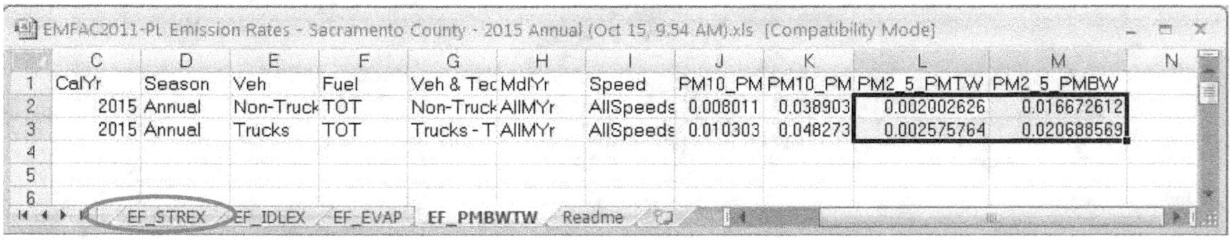
EMFAC2011-PL Emission Rates - Sacramento County - 2015 Annual (Oct 15, 9.54 AM).xls [Compatibility Mode]

	C	D	E	F	G	H	I	J	K	L	M
1	CalYr	Season	Veh	Fuel	Veh & Tec	MdlYr	Speed	PM10_PM	PM10_PM	PM2_5_PMTW	PM2_5_PMBW
2	2015	Annual	Non-Truck	TOT	Non-Truck	AllMYr	AllSpeeds	0.008011	0.038903	0.002002626	0.016672612
3	2015	Annual	Trucks	TOT	Trucks - T	AllMYr	AllSpeeds	0.010303	0.048273	0.002575764	0.020688569

EF_STREX | EF_IDLEX | EF_EVAP | EF_PMBWTW | Readme

These rates are then summed separately for Trucks and Non-Truck categories (shown in Exhibit G-6).

Exhibit G-6. Calculation of Truck and Non-Truck Total PM$_{2.5}$ EF

	Running Exhaust EF	Tire wear EF	Break wear EF	Total PM$_{2.5}$ EF
Non-trucks	0.0022297	0.0020026	0.0166726	**0.020905**
Trucks	0.0229593	0.0025758	0.0206886	**0.046224**

From the calculated Total PM$_{2.5}$ EF, the truck and non-truck rates are then weighted together based on the relative VMT for each vehicle type. In this example, trucks account for 25% of VMT while non-trucks account for 75% of VMT. Exhibit G-7 demonstrates how the EFs are weighted to calculate a single link emission rate.

Exhibit G-7. Calculation of Total PM$_{2.5}$ Link Emission Rate

	Total Emission Rate	VMT adjustment	Weighted Emission Rate
Non-trucks	0.020905	0.75	0.0156788
Trucks	0.046224	0.25	0.011556
			0.027235

This completes the use of the EMFAC2011-PL tool to determine emissions factors for this project using the simplified approach. The total running link emission factor of 0.027235 grams per vehicle-mile can be now be used in combination with link length and link volume as inputs into the selected air quality model, as discussed in Section 7 of the guidance.

Appendix H:
Example of Using EMFAC2011 to Develop Emission Factors for a Transit Project

H.1 INTRODUCTION

The purpose of this appendix is to illustrate the modeling steps required for users to develop PM idling emission factors for a hypothetical bus terminal project using EMFAC2011. It also shows how to generate emission factors from EMFAC2011 for a project that involves a limited selection of vehicle classes (e.g., urban buses) and an age distribution that differs from the EMFAC2011 defaults.[1] Because the project age distribution differs from the EMFAC2011 defaults, use of the simplified approach and EMFAC2011-PL tool is not appropriate. Instead, the detailed approach described in Section 5.6 of the guidance will be used.

This example uses the "Emfac" mode in EMFAC2011-LDV to generate grams per vehicle-hour (g/veh-hr) emission factors stored in the "Summary Rate" output file (.rts file) suitable for use in the AERMOD air quality model. This example does not include the subsequent air quality modeling; refer to Appendix F for an example of how to run AERMOD for a transit project for PM hot-spot analyses.

The assessment of a bus terminal or other non-highway project can involve modeling two different categories of emissions: (1) the idle and/or start emissions at the project site, and (2) the running exhaust emissions on the links approaching and departing the project site. As discussed in Section 5.7.4, EMFAC2011-LVD allows users to generate emission factors for all of these in a single run. This appendix walks through the steps to model idle emissions for this hypothetical project. Users will be able to generate idle emission factors in a single EMFAC2011-LDV model run; multiple calendar years can also be handled within one model run. As described in the main body of this section, each run will be specific to either PM_{10} or $PM_{2.5}$; however, this example is applicable to both. This example is intended to help project sponsors understand how to create representative idle emission factors based on the best available information supplied by EMFAC2011, thus providing an example of how users may have to adapt the information in EMFAC2011 to their individual project circumstances.

To estimate idle emissions at a terminal project, the main task will involve modifying the default vehicle populations and VMT distribution, by vehicle, fuel, and age distribution embedded in EMFAC2011 to reflect the project-specific bus fleet.

---

[1] This is a highly simplified example showing how to employ EMFAC2011 to calculate idle emission factors for use in air quality modeling. An actual project would be expected to be significantly more complex.

H.2 PROJECT CHARACTERISTICS

A PM_{10} hot-spot analysis is conducted for a planned bus terminal project in Sacramento County, California. The project's first full year of operation is assumed to be the year 2013. Through the interagency consultation process, it is determined that 2015 should be the analysis year (based on the project's emissions and background concentrations). The PM analysis is focused on idle emissions from buses operated in the terminal. Additionally, all buses in this example operate using diesel fuel and are ten years old (age 10).

It is determined that the appropriate EMFAC2011 vehicle category for the urban transit buses included in the project is "UBUS-DSL," which is a type found in the EMFAC2011-LDV module (see Section 5.6.2 of the guidance). Therefore, we will be applying the EMFAC2011-LDV procedure described in Section 5.7 of the guidance.

H.3 PREPARING EMFAC2011 BASIC INPUTS

Based on the project characteristics, basic inputs and default settings in EMFAC2011-LDV are first specified (see Exhibit H-1). These basic inputs are similar to those specified for highway projects. To generate idle emission factors for urban transit buses (UBUS-DSL) from EMFAC2011-LDV, a speed bin of 5 mph must be selected in the EMFAC2011-LDV interface.

Exhibit H-1. Basic Inputs in EMFAC2011-LDV for the Hypothetical Highway Project

Step	Input Category	Input Data	Note
1	Geographic Area	County → Sacramento	Select from drop-down list
	Calculation Method	Use Average	Default (not visible in the EMFAC2011-LDV user interface)
2	Calendar Years	2015	Select from drop-down list
3	Season or Month	Annual	Select from drop-down list
4	Scenario Title	Use default	Define default title in the EMFAC2011-LDV user interface
5	Model Years	Use default	Include all model years
6	Vehicle Classes	Use default	Include all vehicle classes
7	I/M Program Schedule	Use default	Include all pre-defined I/M program parameters
8	Temperature	60F	Delete all default temperature bins and input 60
9	Relative Humidity	70%RH	Delete all default relative humidity bins and input 70
10	Speed	Use default	Include speed bin of 5 mph
11	Emfac Rate Files	Summary Rates (RTS)	Select from EMFAC2011-LDV user interface
12	Output Particulate	PM_{10}	Select from EMFAC2011-LDV user interface

H.4 EDITING EMFAC2011-LDV DEFAULT VMT AND POPULATION TO REFLECT PROJECT-SPECIFIC BUS FLEET

To generate idle emission factors that reflect the bus terminal project data, vehicle population and VMT by vehicle class must be modified in the EMFAC2011-LDV user interface. The EMFAC2011 module has data limitations regarding idle emissions: among the available vehicle classes in EMFAC2011-LDV, idle emission factors are available only for the LHDT1, LHDT2, MHDT, HHDT, School Buses, and Other Buses vehicle types. Although EMFAC2011-LDV does not explicitly provide idle emission factors for the "UBUS-DSL" class (the class most typically associated with urban transit buses), as described in Section 5.7.4 of the guidance, the 5 mph emission factors may be used to represent transit buses by multiplying the rate (grams/vehicle-mile) by 5 miles per hour, resulting in a grams/veh-hour rate.

Since the fuel use and age distribution of the bus fleet are known, it is necessary to edit the EMFAC2011-LDV program constants (defaults) to reflect this information. First, VMT "By Vehicle and Fuel" will be edited to reflect entirely diesel Urban Bus operation by changing gasoline Urban Bus VMT to "1" (because "0" will cause an error). Next, Population "By Vehicle and Fuel" will be edited to reflect entirely diesel Urban Bus operation by changing the number of gasoline Urban Buses to "1". Finally, the Population "By Vehicle/Fuel/Age" will be edited to reflect the known Urban Bus age distribution by preserving the number of Urban Buses "age 10", and changing the number of buses of all other ages to "0" (note this must be done by exporting the default age distribution to Excel, as explained in Exhibit H-4).

As shown in Figure H-2, VMT is edited to reflect only diesel operation by Urban Buses. For this example bus terminal, a very low value ("1") is entered into the interface for gasoline Urban Buses to represent the project-specific fuel data.

Exhibit H-2. Changing EMFAC2011-LDV Default VMT to Reflect Project-Specific Fuel Use

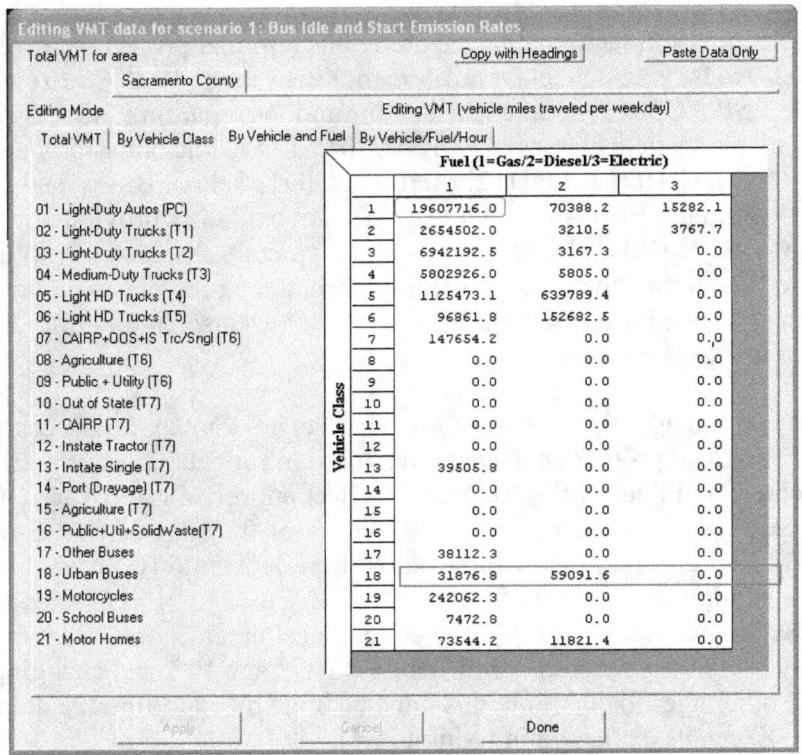

Default EMFAC2011-LDV data before modification

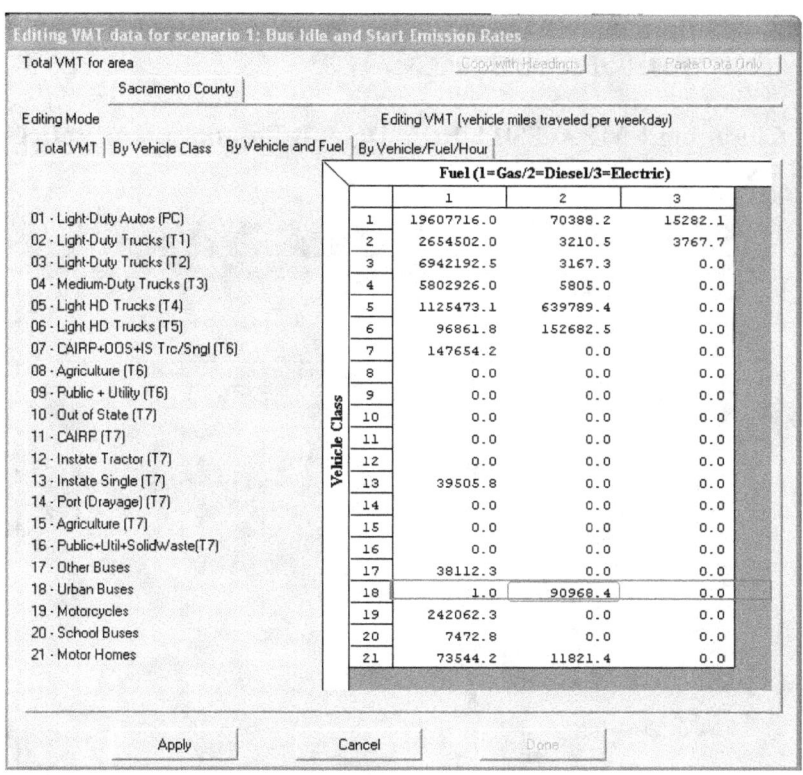

Modified EMFAC2011-LDV data

Next, in Exhibit H-3 the default EMFAC2011-LDV vehicle population is similarly edited to reflect an entirely diesel-fueled bus fleet.

Exhibit H-3. Changing EMFAC2011-LDV Default Population to Reflect Project-Specific Fuel Use

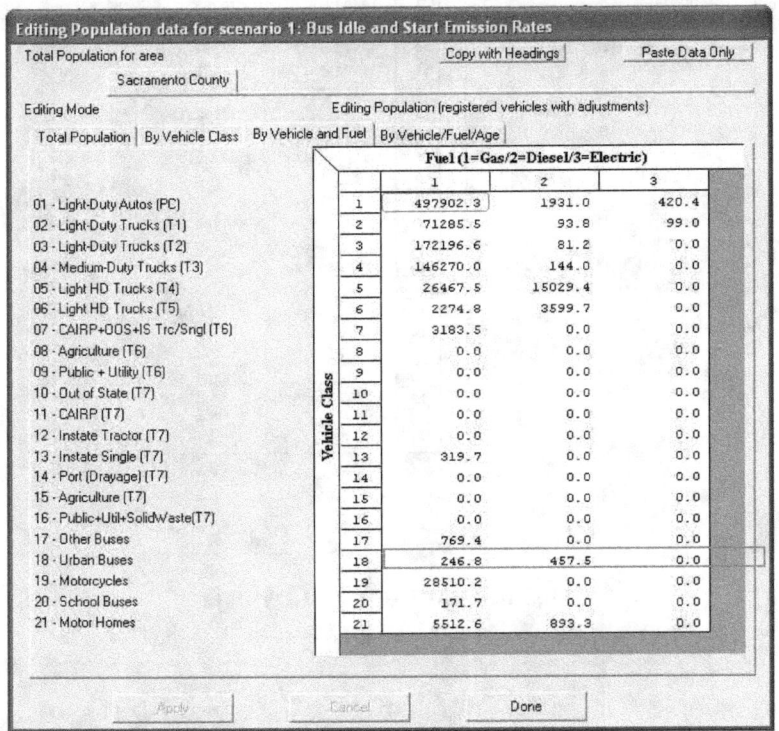

Default EMFAC2011-LDV data before modification

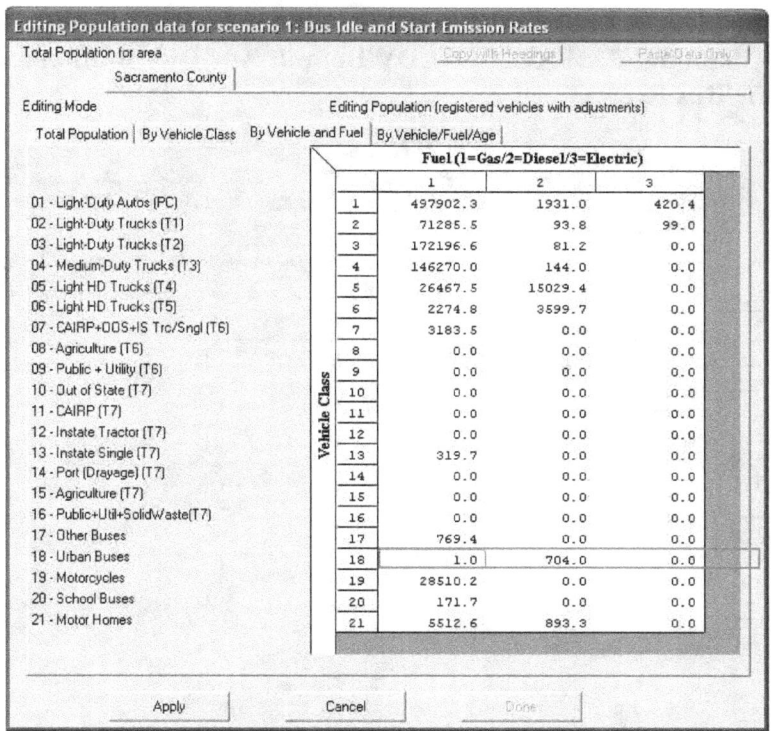

Modified EMFAC2011-LDV data

Finally, in Exhibit H-4, it is necessary to export the default age distribution for modification in Excel. The Urban Bus type has a default age distribution that does not match the project. To change the default, zeros ("0") are entered for all ages except "Age10" to reflect a fleet that is entirely 10 year-old buses. The table is copied and pasted back into the EMFAC2011-LDV module.

Exhibit H-4. Changing EMFAC2011-LDV Default Age Distribution to Reflect Project-Specific Bus Roster

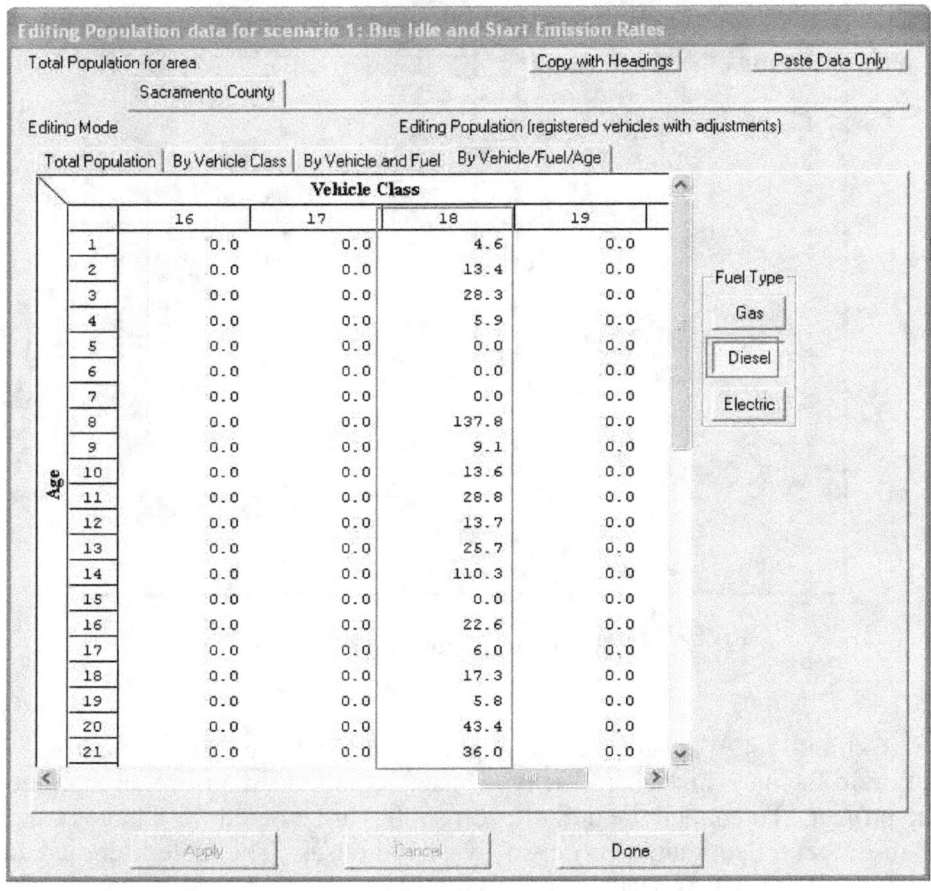

Book1

	A	B	C	D	E	F	G	H	I	J	K
1	Sacramento County Diesel P	Age01	Age02	Age03	Age04	Age05	Age06	Age07	Age08	Age09	Age10
2	01 - Light-Duty Autos (PC)	133.4758	137.1166	152.1877	158.1884	173.6059	179.5487	103.6043	4.97377	0	
3	02 - Light-Duty Trucks (T1)	6.95752	7.152435	7.606674	6.278053	8.160515	5.866775	0	0	0	
4	03 - Light-Duty Trucks (T2)	4.866089	9.060982	10.413	5.814407	4.593681	6.801266	0	0	0	6.659
5	04 - Medium-Duty Trucks (T3	8.03714	7.291153	8.159173	7.372656	7.056864	5.781903	44.25079	3.721883	3.715413	2.745
6	05 - Light HD Trucks (T4)	707.3047	684.5059	650.1201	577.6885	531.6475	465.5589	110.2175	491.8519	599.908	1337.
7	06 - Light HD Trucks (T5)	171.5688	161.188	160.5263	150.7287	129.66	109.6431	53.12713	256.6604	243.032	419.9
8	07 - CAIRP+OOS+IS Trc/Sngl (	0	0	0	0	0	0	0	0	0	
9	08 - Agriculture (T6)	0	0	0	0	0	0	0	0	0	
10	09 - Public + Utility (T6)	0	0	0	0	0	0	0	0	0	
11	10 - Out of State (T7)	0	0	0	0	0	0	0	0	0	
12	11 - CAIRP (T7)	0	0	0	0	0	0	0	0	0	
13	12 - Instate Tractor (T7)	0	0	0	0	0	0	0	0	0	
14	13 - Instate Single (T7)	0	0	0	0	0	0	0	0	0	
15	14 - Port (Drayage) (T7)	0	0	0	0	0	0	0	0	0	
16	15 - Agriculture (T7)	0	0	0	0	0	0	0	0	0	
17	16 - Public+Util+SolidWaste(	0	0	0	0	0	0	0	0	0	
18	17 - Other Buses	0	0	0	0	0	0	0	0	0	
19	18 - Urban Buses	4.634178	13.41263	28.27333	5.890194	0	0	0	137.847	9.056062	13.58
20	19 - Motorcycles	0	0	0	0	0	0	0	0	0	
21	20 - School Buses	0	0	0	0	0	0	0	0	0	
22	21 - Motor Homes	30.27832	28.08069	25.44299	22.05225	19.21754	19.0921	8.860383	44.64226	56.58108	70.4
23											
24											

Sheet1　Sheet2　Sheet3

Default EMFAC2011-LDV age distribution before modification

Book1

	A	B	C	D	E	F	G	H	I	J	K	L	M
1	Sacramen	Age01	Age02	Age03	Age04	Age05	Age06	Age07	Age08	Age09	Age10	Age11	Age12
2	01 - Light-	133.4758	137.1166	152.1877	158.1884	173.6059	179.5487	103.6043	4.97377	0	0	0	
3	02 - Light-	6.95752	7.152435	7.606674	6.278053	8.160515	5.866775	0	0	0	0	0	
4	03 - Light-	4.866089	9.060982	10.413	5.814407	4.593681	6.801266	0	0	0	6.659657	7.168968	
5	04 - Mediu	8.03714	7.291153	8.159173	7.372656	7.056864	5.781903	44.25079	3.721883	3.715413	2.745002	0	1.7756
6	05 - Light I	707.3047	684.5059	650.1201	577.6885	531.6475	465.5589	110.2175	491.8519	599.908	1337.003	1116.968	1365.
7	06 - Light I	171.5688	161.188	160.5263	150.7287	129.66	109.6431	53.12713	256.6604	243.032	419.9518	341.9187	183.
8	07 - CAIRP	0	0	0	0	0	0	0	0	0	0	0	
9	08 - Agricu	0	0	0	0	0	0	0	0	0	0	0	
10	09 - Public	0	0	0	0	0	0	0	0	0	0	0	
11	10 - Out o	0	0	0	0	0	0	0	0	0	0	0	
12	11 - CAIRP	0	0	0	0	0	0	0	0	0	0	0	
13	12 - Instat	0	0	0	0	0	0	0	0	0	0	0	
14	13 - Instat	0	0	0	0	0	0	0	0	0	0	0	
15	14 - Port (I	0	0	0	0	0	0	0	0	0	0	0	
16	15 - Agricu	0	0	0	0	0	0	0	0	0	0	0	
17	16 - Public	0	0	0	0	0	0	0	0	0	0	0	
18	17 - Other	0	0	0	0	0	0	0	0	0	0	0	
19	18 - Urban	0	0	0	0	0	0	0	0	0	704	0	
20	19 - Motor	0	0	0	0	0	0	0	0	0	0	0	
21	20 - Schoo	0	0	0	0	0	0	0	0	0	0	0	
22	21 - Motor	30.27832	28.08069	25.44299	22.05225	19.21754	19.0921	8.860383	44.64226	56.58108	70.4305	59.19204	64.17
23													
24													

Sheet1　Sheet2　Sheet3

Modified age distribution

H-9

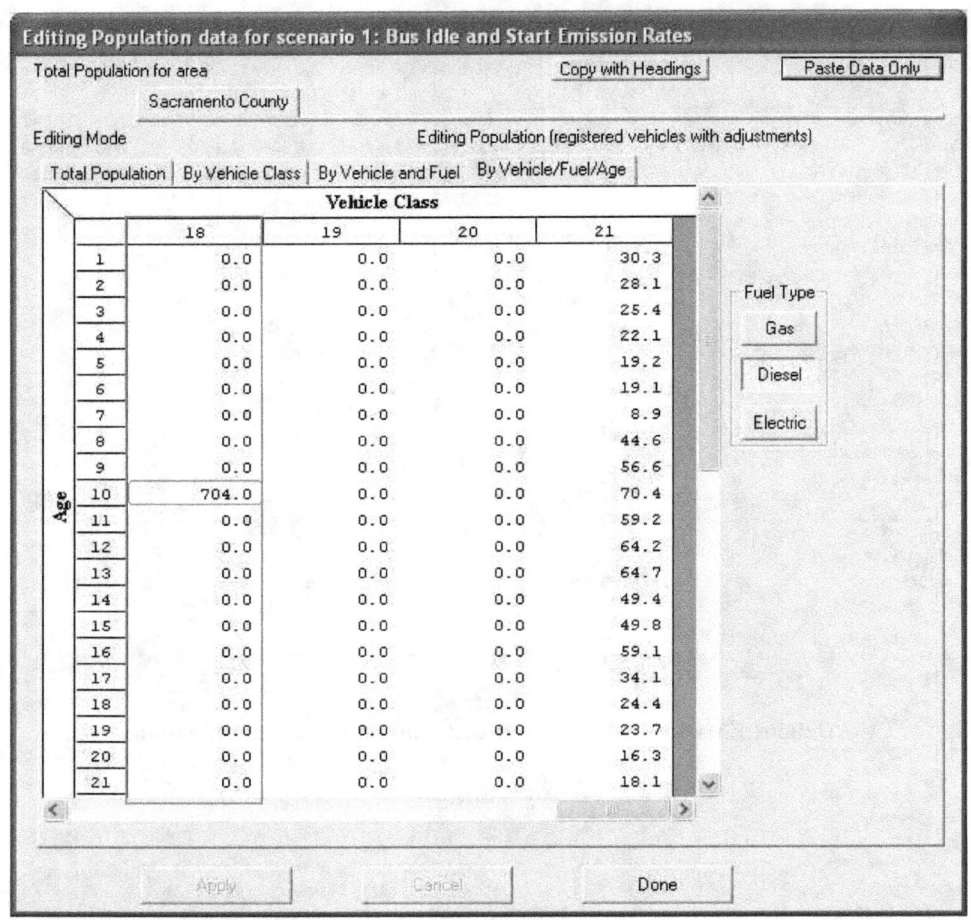

Modified EMFAC2011-LDV data

H.6 PROCESSING IDLE EMISSION FACTORS

Urban Buses ("UBUS") is the vehicle class best representing transit buses in this hypothetical bus terminal project. After the EMFAC2011-LDV run is completed, the project-specific idle exhaust emission factors are presented in Table 1 of the output Summary Rates file (.rts file) as shown in Exhibit H-5.

Exhibit H-5. EMFAC2011-LDV Output

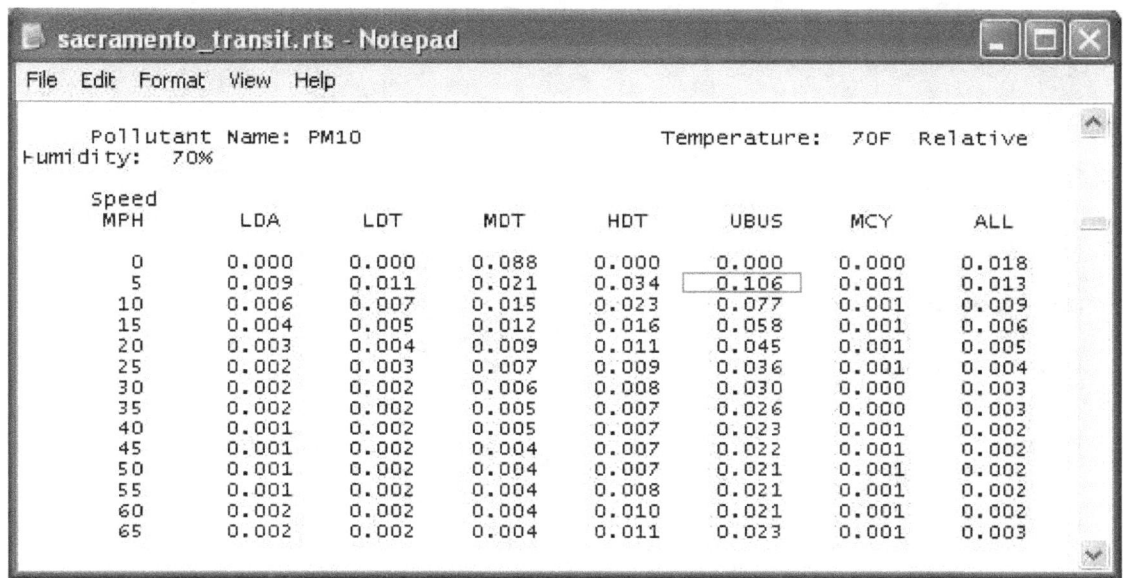

As discussed, the Urban Bus type does not have an explicit idle emission rate. Therefore, the 5 mph emission rate will be used to represent idle operation. As highlighted in Exhibit H-5, the PM_{10} 5 mph exhaust emission factor for the Urban Buses is 0.106 grams/veh-mile. In order to produce a grams/veh-hour emission factor for use in AERMOD, this emission factor (0.106 grams/vehicle-mile) is multiplied by 5 miles per hour. The resulting rate is 0.53 grams/veh-hour. Note that buses typically do not idle for the entire hour, so this rate should be applied to the actual number of bus idle-hours (i.e., [grams/vehicle-hour] x [idling time of each vehicle in fraction of an hour] x [number of vehicles]) expected in the project area to produce an updated grams/hour rate.

This completes the use of EMFAC2011-LDV for determining idle emission factors for this project. The grams/hour idle rate can now be input into AERMOD as discussed in Section 7 of the guidance.

Appendix I:
Estimating Locomotive Emissions

I.1   INTRODUCTION

This appendix describes how to quantify locomotive emissions when they are a component of a transit or freight terminal or otherwise a source in the project area being modeled. Note that state or local air quality agencies may have experience modeling locomotive emissions and therefore could be of assistance when quantifying these emissions for a PM hot-spot analysis.

Generally speaking, locomotive emissions can be estimated in the following manner:

1. Determine where in the project area locomotive emissions should be estimated.

2. Determine when to analyze emissions.

3. Describe the locomotive activity within the project area, including:
    - The locomotives present in the project area (the "locomotive roster"); and
    - The percentage of time each locomotive spends in various throttle settings (the "duty cycle").

4. Calculate locomotive emissions using either:
    - Horsepower rating and load factors, or
    - Fuel consumption data.[1]

The estimated locomotive emission rates that result from this process would then be used for air quality modeling. The interagency consultation process must be used to evaluate and choose the model and associated method and assumptions used for quantifying locomotive emissions for PM hot-spot analyses (40 CFR 93.105(c)(1)(i)).

I.2   DETERMINING WHERE IN THE PROJECT AREA LOCOMOTIVE EMISSIONS SHOULD BE ESTIMATED

Under certain circumstances, it is appropriate to model different locations within the project area as separate sources to characterize differences in locomotive type and/or activity appropriately. This step is analogous to dividing a highway project into links (as described in Sections 4.2 and 5.2 of the guidance) and improves the accuracy of emissions modeling and subsequent air quality modeling. For example, in an intermodal terminal, emissions from a mainline track (which will have a large percentage of higher

---

[1] These are the two methods described in this appendix; others may be possible. See Appendix I.5 for details.

speed operations with little idling) should be estimated separately from the associated passenger or freight terminal (which would be expected to experience low speed operations and significant idling).

The following activities are among those typically undertaken by locomotives and are candidates for being modeled as separate sources if they occur at different locations within the project area:

- Idling within the project area;
- Trains arriving into, or departing from, the project area (e.g., terminal arrival and departure operations);
- Testing, idling, and service movements in maintenance areas or sheds;
- Switching operations;
- Movement of trains passing through, but not stopping in, the project area.

The project area may also be divided into separate sources if it includes several different locomotive rosters (see Appendix I.4.1, below)

I.3   DETERMINING WHEN TO ANALYZE EMISSIONS

The number of hours and days that have to be analyzed depends on the range of activity expected to occur within the project area. For rail projects where activity varies from hour to hour, day to day, and possibly month to month, it is recommended that, at a minimum, project sponsors calculate emissions based on 24 hours of activity for both a typical weekday and weekend day and for four representative quarters of the analysis year when comparing emissions to all $PM_{2.5}$ NAAQS.[2] For projects in areas that violate only the 24-hour PM_{10} or $PM_{2.5}$ NAAQS, the project sponsor may choose to model only one quarter, in appropriate cases. See Section 3.3.4 of the guidance for further information.

These resulting emission rates should be applied to AERMOD and used to calculate design values to compare with the applicable PM NAAQS as described in Sections 7 through 9 of the guidance.

I.4   DESCRIBING THE LOCOMOTIVE ROSTERS AND DUTY CYCLES

Before calculating locomotive emission rates, it is necessary to know what locomotives are present in the locations being analyzed in the project area (see Appendix I.2, above) and what activities these locomotives are undertaking at these locations. This data will impact how emissions are calculated.

---

[2] If there is no difference in activity between weekday and weekend activity, it may not be necessary to examine weekend day activity separately. Similarly, if there is no difference in activity between quarters, emission rates can be determined for one quarter, which can then be used to represent every quarter of the analysis year.

I.4.1   Locomotive rosters

Because emissions can vary significantly depending on a locomotive's make, model, engine, and year of engine manufacture (or re-manufacture), it is important to know what locomotives are expected to be operating within the project area. Project sponsors should develop a "locomotive roster" (i.e., a list of each locomotive's make, model, engine, and year) for the locomotives that will be operating within the specific project area being analyzed. The more detailed the locomotive roster, the more accurate the estimated emissions will be.

In some cases, it will be necessary to develop more than one locomotive roster to reflect the operations in the project area accurately (for example, switcher locomotives may be confined to one portion of a facility and therefore may be represented by their own roster). In these situations, users should model areas with different rosters as separate sources to account for the variability in emissions (see Appendix I.2).

I.4.2   Locomotive duty cycles

Diesel locomotive engine power is controlled by "notched" throttles; idling, braking, and moving the locomotive is conducted by placing the throttle in one of several available "notch settings."[3] A locomotive's "duty cycle" is a description of how much time, on average, the locomotive spends in each notch setting when operating. Project sponsors should use the latest locally-generated or project-specific duty cycles whenever possible; this information may be available from local railway authorities or the state or local air agency.[4] The default duty cycles for line-haul and switch locomotives, found in Tables 1 and 2 of 40 CFR 1033.530 (EPA's regulations on controlling emissions from locomotives), should be used only if they adequately represent the locomotives that will be present in the project area and no local or project-specific duty cycles are available.

I.5   CALCULATING LOCOMOTIVE EMISSIONS

Once a project's locomotive rosters and respective duty cycles have been determined, locomotive emissions can then be calculated for each part of the project area using either (1) horsepower rating and load factors, or (2) fuel consumption data. These two methods are summarized below. Unless otherwise determined through consultation, only one method should be used for a given project.

---

[3] A diesel locomotive typically has eight notch settings for movement (run notches), in addition to one or more idle or dynamic brake notch settings. Dynamic braking is when the locomotive engine, rather than the brake, is used to control speed.

[4] The state or local air agency may have previously developed locally-appropriate duty cycles for emissions inventory purposes.

I.5.1 Finding emission factors

Regardless of method chosen, locomotive emissions factors will be needed for the analysis. Locomotive emission factors depend on the type of engine, the power rating of the locomotive (engine horsepower), and the year of engine manufacture (or re-manufacture). Default PM_{10} emission factors for line-haul and switch locomotives can be obtained from Tables 1 and 2 of EPA's "Emission Factors for Locomotives," EPA-420-F-09-025 (April 2009).[5] These PM_{10} emission factors are in grams/horsepower-hour and can easily be converted to $PM_{2.5}$ emission factors. However, these are simply default values; locomotive-specific data may be available from manufacturers and should be used whenever possible. In addition, see Appendix I.5.4 for other variables that must be considered when determining the appropriate locomotive emission factors.

Note that the default locomotive emission factors promulgated by EPA may change over time as new information becomes available. The April 2009 guidance cited above contains the latest emission factors as of this writing. Project sponsors should consult the EPA's website at: www.epa.gov/otaq/locomotives.htm for the latest locomotive default emission factors and related guidance.

I.5.2 Calculating emissions using horsepower rating and load factors

One way locomotive emissions can be calculated is to use $PM_{2.5}$ or PM_{10} locomotive emission factors, the horsepower rating of the engines found on the locomotive roster, and engine load factors (which are calculated from the duty cycle).

Calculating Engine Load Factors

The horsepower of the locomotive engines, including the horsepower used in each notch setting, should be available from the rail operator or locomotive manufacturer. Locomotive duty cycle data (see Appendix I.4.2) can then be used to determine how much time each locomotive spends in each notch setting, including braking and idling. An engine's "load factor" is the percent of maximum available horsepower it uses over the course of its duty cycle. In other words, a load factor is the weighted average power used by the locomotive divided by the engine's maximum rated power.[6] Load factors can be calculated by summing the actual horsepower-hours of work generated by the engine in a given period of time and dividing it by the engine's maximum horsepower and the hours during which the engine was being used, with the result expressed as a percentage. For example, if a 4000 hp engine spends one hour at full power (generating 4000 hp-hrs) and one hour at 50 percent power (generating 2000 hp-hrs), its load factor would be 75

---

[5] Table 1 of EPA's April 2009 document includes default emission factors for higher power cycles representative of general line-haul operation; Table 2 includes emission factors for lower power cycles used for switching operations. The April 2009 document also includes information on how to convert PM_{10} emission factors for $PM_{2.5}$ purposes. Note that Table 6 (PM_{10} Emission Factors) should not be used for PM hot-spot analyses, since these factors are national fleet averages rather than emission factors for any specific project.

[6] "Weighted average power" in this case is the average power used by the locomotive weighted by the time spent in each notch, as explained further below.

percent (6000 hp-hrs ÷ 4000 hp ÷ 2 hrs). Note that, in this example, it would be equivalent to calculate the load factor using the percent power values instead: ((100% * 1 hr) + (50% * 1 hr) ÷ 2 hrs = 75%). To simplify emission factor calculations, it is recommended that locomotive activity be generalized into the operational categories of "moving" and "idling," with separate load factors calculated for each.

An engine's load factor is calculated by completing the following steps:

Step 1. Determine the number of notch settings the engine being analyzed has and the horsepower used by the engine in each notch setting.[7] Alternatively, as described above, the percent of maximum power available in each notch could instead be used.

Step 2. Identify the percentage of time the locomotive being analyzed spends in each notch setting based on its duty cycle (see Appendix I.4.2).

Step 3. To make emission rate calculations easier, it is useful to calculate two separate load factors for an engine: one for when the locomotive is idling and one for when it is moving.[8] Therefore, the percentage of time the locomotive spends in each notch (from Step 2) needs to be adjusted so that all idling and all moving notches are considered separately. For example, if a locomotive has just one idle notch setting, it spends 100% of its idling time in that setting, even if it only idles during part of its duty cycle. While calculating the time spent idling will usually be simple, for the non-idle (moving) notch settings some additional adjustment to the locomotive's duty cycle percentages will be required to determine the time spent in each moving notch as a fraction of total time spent moving, disregarding any time spent idling.

For example, say a locomotive spends 30% of its time idling and 70% of its time moving over the course of its duty cycle and that 15% of this total time (idling and moving together) is spent in notch 2. When calculating the moving load factor, this percentage needs to be adjusted to determine what fraction of just the 70% of time spent moving is spent in notch 2. In this example, 15% of the total duty cycle spent in notch 2 would equal 21.4% (15% * 100% ÷ 70%) of the locomotive's time when it is not at idle; that is, whenever it is moving, this locomotive spends 21.4% of its time in notch 2. This calculation is repeated for each moving notch setting. The result will be the fraction of time spent in each notch when considering idle and moving modes of operation separately.

Step 4. The next step is to calculate what fraction of maximum available horsepower is being used based on the time spent in each notch setting as was calculated in Step 3. This is determined by summing the product of the percentage of time spent in each notch (calculated in Step 3) by the horsepower generated by the engine at that notch setting (determined in Step 1). For example, if the locomotive with a rated engine power of

---

[7] For locomotives that are equipped with multiple dynamic braking notches and/or multiple idle notches, it may be necessary to assume a single dynamic braking notch and a single idle notch, depending on what information is available about the particular engine.

[8] In this case, "moving" refers to all non-idle notch settings: that is, dynamic braking and all run notches.

3000 hp spends 21.4% of its moving time in notch 2 and 78.6% of its moving time in notch 6, and is known to generate 500 hp while in notch 2 and 2000 hp while in notch 6, then its weighted average power would be 1679 hp (107 hp (500 hp * 0.214) + 1572 hp (2000 hp * 0.786) = 1679 hp).

Step 5. The final step is to determine the load factors. This is done by dividing the weighted average horsepower (calculated in Step 4) by the maximum engine horsepower. For idling, this should be relatively simple. For example, if there is one idle notch setting and it is known that a 4000 hp engine uses 20 hp when in its idle notch, then its idle load factor will be 0.5% (20 hp ÷ 4000 hp). To determine the load factor for all power notches, the weighted horsepower calculated in Step 4 should be divided by the total engine horsepower. For example, if the same 4000 hp engine is determined to use an average of 1800 hp while in motion (as determined by adjusting the horsepower by the time spent in each "moving" notch setting in Step 4), then the moving load factor would be 45% (1800 hp ÷ 4000 hp).

The resulting idling and moving load factors represent the average amount of the total engine horsepower the locomotive is using when idling and moving, respectfully. These load factors can then be used to modify PM emission factors and generate emission rates as described below.

Generating Emission Rates Based on Load Factors

As noted above, EPA's "Emission Factors for Locomotives" provides emission factors in grams/brake horsepower-hour. This will also likely be the case with any specific emission factors obtained from manufacturer's specifications. These units can be converted into grams/second (g/s) emission rates by using the load factor on the engines and the time spent in each operating mode, as described below.

The first step is to adjust the PM emission factors to reflect how the engine will actually be operating.[9] This is done by multiplying the appropriate PM emission factor by the idling and moving load factors calculated for that particular engine.[10] Next, to determine the emission rate, this adjusted emission factor is further multiplied by the amount of time the locomotive spends idling and moving while in the project area.[11]

For example, if the PM emission factor known to be 0.18 g/bhp-hr, the engine being analyzed has an idling load factor of 0.5%, and the locomotive is anticipated to idle 24 minutes per hour in the project area, then the resulting emission rate would be 0.035 grams/hour (0.18 g/bhp-hr * 0.5% * 0.4 hours).

---

[9] Because combustion characteristics of an engine vary by throttle notch position, it is appropriate to adjust the emission factor to reflect the average horsepower actually being used by the engine.

[10] Project sponsors are reminded to check www.epa.gov/otaq/locomotives.htm to ensure the latest default emission factors for idle and moving emissions are being used.

[11] Note that this may or may not match up with the idle and moving time as described by the duty cycle used to calculate the load factors, depending on how project-specific that duty cycle is.

Emission rates need to be converted into g/s for use by AERMOD, as described further in Sections 7 through 9 of the guidance. These calculations should be repeated until the entire locomotive roster is represented in each part of the project area being analyzed.

Appendix I.7 provides an example of calculating g/s locomotive emission rates using this methodology.

I.5.3   Calculating emissions using fuel consumption data

Another method to calculate locomotive emissions involves using fuel consumption data. Chapter 6.3 of EPA's "Procedure for Emission Inventory Preparation -- Volume IV: Mobile Sources" (reference information provided in Appendix I.6, below) is a useful reference and should be consulted when using this method.

Note that, for this method, it may be useful to scale down data already available to the project sponsor. For example, if rail car miles/fuel consumption is known for trains operating in situations identical to those being estimated in the project area, this data can be used to estimate fuel consumption rates for a defined track length within the project area.

Calculating Average Fuel Consumption

Locomotive fuel consumption is specific to a particular locomotive engine and the throttle (notch) setting it is using. Data on the fuel consumption of various engines at different notch settings can often be obtained from the locomotive or engine manufacturer's specifications. When only partial data is available (e.g., only data for the lowest and highest notch settings are known), interpolation combined with best available engineering judgment can be used to determine fuel consumption at the intermediate notch settings.

A locomotive's average fuel consumption can be calculated by determining how long each locomotive is expected to spend in each notch setting based on its duty cycle (see Appendix I.4.2). This data can be aggregated to generate an average fuel consumption rate for each locomotive type. See Chapter 6.3 of Volume IV for details on how to generate this data based on a specific locomotive roster and duty cycle.

Once the average fuel consumption rates have been determined, they should be multiplied by the appropriate emission factors to determine a composite average hourly emission rate for each engine in the roster. Since the objective is to determine an average fuel consumption rate for the entire locomotive roster, this calculation should be repeated for each engine on the roster at each location analyzed.

If several individual sources will be modeled at different sections of the project area as described in Appendix I.2, train schedule data should be consulted to determine the hours of operation of each locomotive within each section of the project area. Hourly emission rates per locomotive should then be multiplied by the number of hours the locomotive is operating, for each hour of the day in each section of the project area to provide average hourly emission rates for each section of the project. These should then be converted to grams/second for use in AERMOD, as described further in Sections 7 through 9 of the guidance.

Examples of calculating locomotive emissions using this method can be found in Chapter 6 of Volume IV.

I.5.4 Factors influencing locomotive emissions and emission factors

The following considerations will influence locomotive emissions regardless of the method used and should be examined when determining how to characterize locomotives for emissions modeling or when choosing the appropriate emission factors:

- Project sponsors should be aware of the emission reductions that would result from remanufacturing existing locomotives (or replacing existing locomotives with new locomotives) that meet EPA's Tier 3 or Tier 4 emission standards when they become available. The requirements that apply to existing and new locomotives were addressed in EPA's 2008 rulemaking entitled "Control of Emissions of Air Pollution from Locomotive Engines and Marine Compression-Ignition Engines Less Than 30 liters Per Cylinder" (73 FR 37095). Beginning in 2012 all locomotives will be required to use ultra-low sulfur diesel fuel (69 FR 38958). Additionally, when existing locomotives are remanufactured, certified remanufacture systems will have to be installed to reduce emissions. Beginning in 2011, new locomotives must meet tighter Tier 3 emission standards. Finally, beginning in 2015 even more stringent Tier 4 emission standards for new locomotives will begin to be phased in.

- For locomotives manufactured before 2005, a given locomotive may be in one of three possible configurations, depending on when it was last remanufactured: (1) uncertified; (2) certified to the standards in 40 CFR Part 92; or (3) certified to the standards in 40 CFR Part 1033. Each of these configurations should be treated as a separate locomotive type when conducting a PM hot-spot analysis.

- Emissions from locomotives certified to meet Family Emission Limits (FELs) may differ from the emission standard identified on the engine's Emission Control Information label. Rail operators will know if their locomotives participate in this program. Any locomotives in the project area participating in this program should be identified so that the actual emissions from the particular locomotives being analyzed are considered in the analysis, rather than the family emissions level listed on their FEL labels.

I.6   AVAILABLE RESOURCES

These resources and websites should be checked prior to beginning any PM hot-spot analysis to ensure that the latest data (such as emission factors) are being used:

- "Emission Factors for Locomotives," EPA-420-F-09-025 (April 2009). Available online at: www.epa.gov/otaq/locomotives.htm.

- Chapter 6 of "Procedure for Emission Inventory Preparation - Volume IV: Mobile Sources." Available online at: www.epa.gov/OMS/invntory/r92009.pdf. Note that, as of this writing, the emission factors listed in Volume IV have been superseded by the April 2009 publication listed above for locomotives certified to meet current EPA standards.[12]

- "Control of Emissions from Idling Locomotives," EPA-420-F-08-014, March 2008. Available online at: www.epa.gov/otaq/regs/nonroad/locomotv/420f08014.htm.

- See Section 10 of the guidance for additional information regarding potential locomotive emission control measures.

I.7   EXAMPLE OF CALCULATING LOCOMOTIVE EMISSION RATES USING HORSEPOWER RATING AND LOAD FACTOR ESTIMATES

The following example demonstrates how to estimate locomotive emissions using the engine horsepower rating/load factor method described in Appendix I.5.2.

The hypothetical proposed project in this example includes the construction of an intermodal terminal in an area that is designated as nonattainment for both the 1997 annual $PM_{2.5}$ NAAQS and the 2006 24-hour $PM_{2.5}$ NAAQS. The terminal in this example is to be completed and operational in 2013. The hot-spot analysis is performed for 2015, because it is determined through interagency consultation that this will be the year of peak emissions, when considering the project's emissions and the other emissions in the project area.

In this example, the operational schedule anticipates that 32 locomotives will be in the project area over a 24-hour period, with 16 locomotives in the project area during the peak hour. Based on the schedule, it is further determined that while in the project area each train will spend 540 seconds idling and 76 seconds moving.

---

[12] Although the emission factors have been superseded, the remainder of the Volume IV guidance remains in effect.

The locomotive $PM_{2.5}$ emissions are calculated based on horsepower rating and load factors.

I.7.1 Calculate idle and moving load factors

As described in I.5.2, the project sponsor uses a series of steps to calculate load factors. These steps are described below and the results from each step are shown in table form in Exhibit I-1.

Step 1: The project sponsor first needs some information about the locomotives expected to be operating at the terminal in the analysis year.

For each locomotive, the horsepower used by the locomotive in each notch setting as well as under dynamic braking and at idle must be determined. For the purpose of this example it is assumed that all of the locomotives that will serve this terminal are very similar: all use the same horsepower under each of operating conditions, and all have only one idle and dynamic braking notch setting. The horsepower generated at each notch setting is obtained from the engine specifications (see second column of Exhibit I-1). In this case, the rated engine horsepower is 4000 hp (generated at notch 8).

Step 2: The next step is to determine the average amount of time that the locomotives spend in each notch and expressing the results as a percentage of the locomotive's total operating time. In this example, it is determined that, based on their duty cycle, the locomotives that will service this terminal spend 38% of their time idling and 62% of their time in motion in one of the eight run notch settings or under dynamic braking. The percentage of time spent in each notch is shown in the third column of Exhibit I-1.

Step 3: To make emission factor calculations easier, it is decided to calculate separate idling and moving load factors. The next step, then, is for the project sponsor to calculate the actual percentage of time that the locomotives spend in each notch, treating idling and moving time separately. This is done by excluding the time spent idling and recalculating the percentage of time spent in the other notches (i.e., dynamic braking and each of the eight notch settings) so that the total time spent in non-idle notches adds to 100%. The results are shown in the fourth column of Exhibit I-1.

Step 4: The next step is to calculate the weighted average horsepower for this engine using the horsepower generated in each notch and the percentage of time spent in each notch as adjusted in Step 3. For locomotives that are idling, this is simply the horsepower used at idle. For the other notches, the actual horsepower for each notch is determined by multiplying the horsepower generated in a given notch (determined in Step 1) by the actual percentage of time that the locomotive is in that notch, as adjusted (calculated in Step 3). The results are shown in the fifth column of Exhibit I-1.

Step 5: The final step in this part of the analysis is to determine the idle and moving load factors. The idle load factor is just the horsepower generated at idle divided by the maximum engine horsepower, with the result expressed as a percentage. To determine the moving load factor, the weighted average horsepower for all non-idle notches (calculated in Step 4) is divided by the maximum engine horsepower, with the result expressed as a percentage. The final column of Exhibit I-1 shows the results of these calculations, with the idling and moving load factors highlighted.

Exhibit I-1. Calculating Locomotive Load Factors

Notch Setting	Step 1: Horsepower (hp) used in notch	Step 2: Average % time spent in notch	Step 3: Reweighted time spent in each notch (adjusted so that non-idle notches add to 100%)	Step 4: Time-weighted hp used, based on time spent in notch	Step 5: Load factors (idle and moving)
Idling load factor:					
Idle	14	38.0%	100.0%	14.0	0.4%
Moving load factor:					
Dynamic Brake	136	12.5%	20.2%	27.5	
1	224	6.5%	10.5%	23.5	
2	484	6.5%	10.5%	50.8	
3	984	5.2%	8.4%	82.7	
4	1149	4.4%	7.1%	81.6	
5	1766	3.8%	6.1%	107.8	
6	2518	3.9%	6.3%	158.6	
7	3373	3.0%	4.8%	161.9	
8	4,000	16.2%	26.1%	1,044.0	
Total		62.0%	100.0%	1,752.4	43.8%

I.7.2   Using the load factors to calculate idle and moving emission rates

Now that the idle and moving load factors have been determined, the gram/second (g/s) emission rates can be calculated for the idling and moving locomotives.

First, the project sponsor would determine how many locomotives are projected to be idling and how many are projected to be in motion during the peak hour of operation and over a 24-hour period. As previously noted, it is anticipated that 32 locomotives will be in the project area over a 24-hour period, with 16 locomotives in the project area during the peak hour. It was further determined that, while in the project area, each train will spend 540 seconds idling and 76 seconds moving.

For the purpose of this example, it has been assumed that each locomotive idles for the same amount of time and is in motion for the same amount of time. Note that, in this case, the number of locomotives considered "moving" will be double the actual number of locomotives present in order to account for the fact that each locomotive moves twice through the project area (as it arrives and departs the terminal).

Next, the project sponsor would determine the $PM_{2.5}$ emission factor to be used in this analysis for 2015. These emission factors can be determined from the EPA guidance titled "Emission Factors for Locomotives."

Table 1 of "Emission Factors for Locomotives" presents PM_{10} emission factors in terms of grams/brake horsepower-hour (g/bhp-hr) for line haul locomotives that are typically used by commuter railroads. Emission factors are presented for uncontrolled locomotives, locomotives manufactured to meet Tier 0 through Tier 4 emission standards, and locomotives remanufactured to meet more stringent emission standards. It's important to determine the composition of the fleet of locomotives that will use the terminal in the year that is being analyzed so that the emission factors in Table 1 can be used in the calculations. This information would be available from the railway operator.

In this example, we are assuming that all of the locomotives meet the Tier 2 emission standard. However, an actual PM hot-spot analysis would likely have a fleet of locomotives that meets a combination of these emission standards. The calculations shown below would have to be repeated for each different standard that applies to the locomotives in the fleet.

The final step in these calculations is to use the information shown in Exhibit I-1 and the other project data collected to calculate the $PM_{2.5}$ emission rates for idling and moving locomotives during both the peak hour and over a 24-hour basis.[13]

Calculating Peak Hour Idling Emissions

The following calculation would be used to determine the idling emission rate during the peak hour of operation:[14]

$PM_{2.5}$ Emission Rate = (16 trains/hr) * (1 hr/3,600 s) * (540 s/train) * (4,000 hp) *
(0.004) * (0.18 g/bhp-hr) * (1 hr/3,600 s) * (0.97)
$PM_{2.5}$ Emission Rate = 0.0019 g/s

Where:
- Trains per hour = 16 (number of trains present in peak hour)
- Idle time per train = 540 s (from anticipated schedule)
- Locomotive horsepower = 4,000 hp (from engine specifications)
- Idle load factor = 0.004 (0.4%, calculated in Exhibit I-1)

---

[13] Peak hour emission rates will not be necessary for all analyses; however, for certain projects that involve very detailed air quality modeling analyses, peak hour emission rates may be necessary to more accurately reflect the contribution of locomotive emissions to air quality concentrations in the project area.

[14] Note that, for the calculations shown here, any units expressed in hours or days need to be converted to seconds since a g/s emission rate is required for AERMOD.

- Tier 2 Locomotive Emission Factor = 0.18 g/bhp-hr (from "Emission Factors for Locomotives")
- Ratio of $PM_{2.5}$ to PM_{10} = 0.97 (from "Emission Factors for Locomotives")

<u>Calculating 24-hour Moving Emissions</u>

Similarly, the following equation would be used to calculate the moving emission rate for the 24-hour period:

$PM_{2.5}$ Emission Rate = (64 trains/day) * (76 s/train) * (1 day/86,400 s) * (4,000 hp) * (0.438) * (0.18 g/bhp-hr) * (1hr/3,600 s) * (0.97)

$PM_{2.5}$ Emission Rate = 0.0048 g/s

Where:
- Trains per day = 64 (double the actual number of trains present over 24 hours to account for each train moving twice through the project area)
- Moving time per train = 76 s (from anticipated schedule)
- Locomotive horsepower = 4,000 hp (from engine specifications)
- Moving load factor = 0.438 (43.8%, calculated in Exhibit I-1)
- Tier 2 Locomotive Emission Factor = 0.18 g/bhp-hr (from "Emission Factors for Locomotives")
- Ratio of $PM_{2.5}$ to PM_{10} = 0.97 (from "Emission Factors for Locomotives")

A summary of the variables used in the above equations and the resulting emission rates can be found in Exhibit I-2, below.

Exhibit I-2. $PM_{2.5}$ Locomotive Emission Rates

Operational Mode	Number of Locomotives		Time/ Train	$PM_{2.5}$ Emission Factor	Calculated Peak Hour Emission Rate	Calculated 24-hour Emission Rate
	Peak hour	24 hours	(s)	(g/bhp-hr)	(g/s)	(g/s)
Idle	16	32	540	0.18	0.0019	0.00016
Moving	32	64	76	0.18	0.057	0.0048

These peak and 24-hour emission rates can now be used in air quality modeling for the project area, as described in Sections 7 through 9 of the guidance.

Note that, since this area is designated as nonattainment for both the 1997 annual $PM_{2.5}$ NAAQS and the 2006 24-hour $PM_{2.5}$ NAAQS, the results of the analysis will be compared to both NAAQS (see Section 3.3.4 of the guidance). Since the area is in nonattainment of the annual $PM_{2.5}$ NAAQS, all four quarters will need to be included in the analysis to estimate a year's worth of emissions. If there is no change in locomotive activity across quarters, the emission rates calculated here could be used for each quarter of the year (see Appendix I.3).

Appendix J:
Additional Reference Information on Air Quality Models and Data Inputs

J.1 INTRODUCTION

This appendix supplements Section 7's discussion of air quality models. Specifically, this appendix describes how to configure AERMOD and CAL3QHCR for PM hot-spot analysis modeling, as well as additional information on handling the data required to run the models for these analyses. This appendix is not intended to replace the user guides for air quality models, but discuss specific model inputs, keywords, and formats for PM hot-spot modeling. This appendix is organized so that it references the appropriate discussions in Section 7 of the guidance.

J.2 SELECTING AN APPROPRIATE AIR QUALITY MODEL

The following discussion supplements Section 7.3 of the guidance and describes how to appropriately configure AERMOD and CAL3QHCR when completing a PM hot-spot analysis. Users should also refer to the model user guides, as appropriate.

J.2.1 Using AERMOD for PM hot-spot analyses

There are no specific commands unique to transportation projects that are necessary when using AERMOD. By default, AERMOD produces output for particulate matter in units of micrograms per cubic meter of air ($\mu g/m^3$). All source types in AERMOD require that emissions are specified in terms of emissions per unit time, although AREA-type sources also require specification of emissions per unit time per unit area. AERMOD has no specific traffic queuing mechanisms. Emissions output from MOVES, EMFAC, AP-42, and other types of methods should be formatted as described in the AERMOD User Guide.[1]

J.2.2 Using CAL3QHCR for PM hot-spot analyses

CAL3QHCR is an extension of the CAL3QHC model that allows the processing of a full year of hourly meteorological data, the varying of traffic-related inputs by hour of the week, and calculation of long-term average concentrations. It also will display the five highest concentration days for the time period being modeled. Emissions output from MOVES, EMFAC, AP-42, and other emission methods should be formatted as described

---

[1] Extensive documentation is available describing the various components of AERMOD, including user guides, model formulation, and evaluation papers. See EPA's SCRAM website for AERMOD documentation: www.epa.gov/scram001/dispersion_prefrec.htm#aermod.

in the CAL3QHCR User Guide.[2] In addition, the following guidance is provided when using CAL3QHCR for a PM hot-spot analysis:

Specifying the Right Pollutant

When using CAL3QHCR for PM hot-spot analyses, the MODE keyword must be used to specify analyses for PM so that concentrations are described in micrograms per cubic meter of air ($\mu g/m^3$) rather than parts per million (ppm).

Entering Emission Rates

MOVES emission rates for individual roadway links are based on the Op-Mode distribution associated with each link and are able to include emissions resulting from idling. MOVES-based emission factors that incorporate relevant idling time and other delays should be entered in CAL3QHCR using the EFL keyword. Therefore, within CAL3QHCR, the IDLFAC keyword's emission rates should be set to zero, because the effects of idling are already included within running emissions. (Note that if a non-zero emission rate is used in CAL3QHCR, the model will treat idling emission rates separately from running emission rates.) The same recommendation applies when using emission rates calculated by EMFAC.

Assigning Speeds

Although the user guide for CAL3QHCR specifies that the non-queuing links should be assigned speeds in the absence of delay caused by traffic signals, the user should use speeds that reflect delay when using CAL3QHCR for a hot-spot analysis. Since MOVES emission factors already include the effects of delay (i.e., Op-Mode distributions that are user-specified or internally calculated include the effects of delay), the speeds used in CAL3QHCR links will already reflect the relevant delay on the link over the appropriate averaging time. The same recommendation applies when using EMFAC.

Using the Queuing Algorithm

When applying CAL3QHCR for the analysis of highway and intersection projects, its queuing algorithm should not be used.[3] This includes the CAL3QHCR keywords NLANE, CAVG, RAVG, YFAC, IV, and IDLFAC. As discussed in Sections 4 and 5, idling vehicle emissions should instead be accounted for by properly specifying links for emission analysis and reflecting idling activity in the activity patterns used for MOVES or EMFAC modeling.

---

[2] The CAL3QHCR user guide and other model documentation can be found on EPA's SCRAM website: www.epa.gov/scram001/dispersion_prefrec.htm#cal3qhc.

[3] CAL3QHCR's algorithm for estimating the length of vehicle queues associated with intersections is based on the *1985 Highway Capacity Manual*, which is no longer current. Furthermore, a number of other techniques are now available that can be used to estimate vehicle queuing around intersections.

J.3 CHARACTERIZING EMISSION SOURCES

The following discussion supplements Section 7.4 of the guidance and describes in more detail how to characterize sources in CAL3QHCR and AERMOD, including the physical characteristics, location, and timing of sources. This discussion assumes the user is familiar with handling data in these models, including the use of specific keywords. For additional information, refer to the CAL3QHCR and AERMOD user guides.

J.3.1   Physical characteristics and locations of sources in CAL3QHCR

CAL3QHCR characterizes highway and intersection projects as line sources. The geometry and operational patterns of each roadway link are described using the following variables, which in general may be obtained from engineering diagrams and design plans of the project:[4]

- The coordinates (X, Y) of the endpoints of each link;[5]
- The width of the "highway mixing zone" (see below);
- The type of link ("at grade," "fill," "bridge," or "depressed");
- The height of the roadway relative to the surrounding ground (not to exceed ±10 meters);[6] and
- The hourly flow of traffic (vehicles per hour).

CAL3QHCR treats the area over each roadway link as a "mixing zone" that accounts for the area of turbulent air around the roadway resulting from vehicle-induced turbulence. The width of the mixing zone is an input to the model. Users should specify the width of a link in CAL3QHCR as the width of the traveled way (traffic lanes, not including shoulders) plus three meters on either side. Users should treat divided highways as two separate links. See Section 7.6 of the guidance for more information on placing receptors.

J.3.2   Timing of emissions in CAL3QHCR

The CAL3QHCR User Guide describes two methods for accepting time-varying emissions and traffic data; these are labeled the "Tier I" and "Tier II" approaches.[7]

---

[4] Traffic engineering plans and diagrams may include information such as the number, width, and configuration of lanes, turning channels, intersection dimensions, and ramp curvature, as well as operational estimates such as locations of weave and merge sections and other descriptions of roadway geometry that may be useful for specifying sources.
[5] In CAL3QHCR, the Y-axis is aligned due north.
[6] The CALINE3 dispersion algorithm in CAL3QHCR is sensitive to the height of the road. In particular, the model treats bridges and above-grade "fill" roadways differently. It also handles below-grade roadways with height of less than zero (0) meters as "cut" sections. Information on the topological features of the project site is needed to make such a determination. Note that in the unusual circumstance that a roadway is more than ten meters below grade, CALINE3 has not been evaluated, so CAL3QHCR is not recommended for application. In this case, the relevant EPA Regional Office should be consulted for determination of the most appropriate model.
[7] This nomenclature is unrelated to EPA's motor vehicle emission standards and the design value calculation options described in Section 9 of this guidance.

Project-level PM hot-spot modeling should use the Tier II method, which can accommodate different hourly emission patterns for each day of the week. Most emissions data will not be so detailed, but the Tier II approach can accommodate emissions data similar to that described in Sections 4 and 5 of the guidance. The CAL3QHCR Tier I approach should not be used, as it employs only one hour of emissions and traffic data and therefore cannot accommodate the emissions data required in a PM hot-spot analysis.

Through the IPATRY keyword, CAL3QHCR allows up to seven 24-hour profiles representing hour-specific emission, traffic, and signalization (ETS) data for each day of the week. Depending on the number of MOVES runs, the emission factors should be mapped to the appropriate hours of the day. For example, peak traffic emissions data for each day would be mapped to the CAL3QHCR entry hours corresponding to the relevant times of day (in this case, the morning and afternoon peak traffic periods). If there are more MOVES runs than the minimum specified in the Section 4, they should be modeled and linked to the correct days and hours using IPATRY.

As described in Section 7 of the guidance, the number of CAL3QHCR runs required for a given PM hot-spot analysis will vary based on the amount of meteorological data available.

J.3.3   Physical characteristics and locations of sources in AERMOD

The following discussion gives guidance on how to best characterize a source. AERMOD includes different commands (keywords) for volume, area, and point sources.

<u>Modeling Volume Sources</u>

Many different sources in a project undergoing a PM hot-spot analysis might be modeled as volume sources. Examples include areas designated for truck or bus queuing or idling (e.g., off-network links in MOVES), driveways and pass-throughs in transit or freight terminals, and locomotive emissions.[8] AERMOD can also approximate a highway "line source" using a series of adjacent volume sources (see the AERMOD User Guide for suggestions). Certain nearby sources that have been selected to be modeled may also be appropriately treated as a volume source (see Section 8 of the guidance for more information on considering background concentrations from other sources).

Volume source parameters are entered using the source parameter (SRCPARAM) keyword in the AERMOD input file. This requires the user to provide the following information:
- The emission rate (mass per unit time, such as g/s);
- The initial lateral dispersion coefficient determined from the initial lateral dimension (width) of the volume;

---

[8] See Section 6 and Appendix I for information regarding calculating locomotive emissions.

- The initial vertical dispersion coefficient determined from the initial vertical dimension (height) of the volume; and
- The source release height of the volume source center, (i.e., meters above the ground).

Within AERMOD, the volume source algorithms are applicable to line sources with some initial plume depth (e.g., highways, rail lines).[9] There are three inputs needed to characterize the initial size of a roadway plume:

1. Initial lateral dispersion coefficient (σ_{yo}, *Syinit*). First, estimate the initial lateral dimension (or width) of the volume source. One of the following options can be used:
   a) The average vehicle width plus 6 meters, when modeling a single lane of traffic;
   b) The road width multiplied by 2; or
   c) A set width, such as 10 meters per lane of traffic.

To specify the initial lateral dispersion coefficient (σ_{yo}), referred to as *Syinit* in AERMOD, the AERMOD User Guide recommends dividing the initial width by 2.15. This is to ensure that the overlapping distributions from adjacent volume sources simulate a line source of emissions.

2. Initial vertical dispersion coefficient (σ_{zo}, *Szinit*). First, estimate the initial vertical dimension (height) of the plume for volume sources. A typical approach is to assume it is about 1.7 times the average vehicle height, to account for the effects of vehicle-induced turbulence. For light-duty vehicles, this is about 2.6 meters, using an average vehicle height of 1.53 meters or 5 feet. For heavy-duty vehicles, this is about 6.8 meters, using an average vehicle height of 4.0 meters. Since most road links will consist of a combination of light-duty and heavy-duty traffic, the initial vertical dimension should be a combination of their respective values. There are two options available to estimate initial vertical dimension:
   a) Estimate the initial vertical dimension using an emissions-weighted average. For example, if light-duty and heavy-duty vehicles contribute 40% and 60% of the emissions of a given volume source, respectively, the initial vertical dimension would be (0.4 * 2.6) + (0.6 * 6.8) = 5.1 meters.
   b) Alternatively, the initial vertical dimension may be estimated using a traffic volume weighted approach based on light-duty and heavy-duty vehicle fractions.

The AERMOD User Guide recommends that the initial vertical dispersion coefficient (σ_{zo}), termed *Szinit* in AERMOD, be estimated for a surface-based volume source by dividing the initial vertical dimension by 2.15. For typical light-duty vehicles, this

---

[9] The vehicle-induced turbulence around roadways with moving traffic suggests that prior to transport downwind, a roadway plume has an initial size; that is, the emissions from the tailpipe are stirred because the vehicle is moving and therefore the plume "begins" from a three-dimensional volume, rather than from a point source (the tailpipe).

corresponds to a *Szinit* (σ_{zo}) of 1.2 meters. For typical heavy-duty vehicles, the initial value of *Szinit* (σ_{zo}) is 3.2 meters.

3. Source release height. The source release height (*Relhgt* in AERMOD), which is the height at which wind effectively begins to affect the plume, may be estimated from the midpoint of the initial vertical dimension. For moving light-duty vehicles, this is about 1.3 meters. For moving heavy-duty vehicles, it is 3.4 meters. Since most road links will consist of a combination of light-duty and heavy-duty traffic, the source release height should be a combination of their respective values. There are two options available to estimate source release height:
   a) Estimate using an emissions-weighted average. For a 40% light-duty and 60% heavy-duty emissions share, the source release height would be (0.4 * 1.3) + (0.6 * 3.4) = 2.6 meters.
   b) Alternatively, the source release height may be estimated using a traffic volume weighted approach based on light-duty and heavy-duty VMT.

Another way of dealing with *Syinit*, *Szinit*, and/or *Relhgt* parameters that change as a result of different fractions of light-duty and heavy-duty vehicles is to create two overlapping versions of each roadway source, corresponding to either light-duty and heavy-duty traffic. These two sources could be superimposed in space, but have emission rates and *Syinit*, *Szinit*, and *Relhgt* parameters that are specific to light-duty or heavy-duty traffic.

Also, AERMOD (version dated 09292) allows *Syinit*, *Szinit*, and *Relhgt* to change by hour of the day, which may be considered if the fraction of heavy-duty vehicles is expected to significantly change throughout a day. Users should consult the latest information on AERMOD when starting a PM hot-spot analysis.

Groups of idling vehicles may also be modeled as one or more volume sources. In those cases, the initial dimensions of the source, dispersion coefficients, and release heights should be calculated assuming that the vehicles themselves are inducing no turbulence.

When using adjacent volume sources to represent emissions from a source such as a roadway, a sufficient number of volume sources should be employed to represent a consistent density of emissions for a single link in a MOVES or EMFAC analysis. In addition, when the source-receptor spacing in AERMOD is shorter than the distance between adjacent volume sources, AERMOD may produce aberrant results. In the present version of the model, receptors within a volume source in AERMOD are assigned concentrations of zero. When volume sources are used and publicly-accessible locations are closer to a source than the distance between adjacent volume sources, it is recommended that smaller volume sources be used with shorter spacing between them.

For example, for such a segment along a highway segment, individual lanes might be modeled discretely, rather than using a single volume source for all lanes. This will reduce the spacing between volume sources and increase the quality of results closest to a source. Receptors near area and point sources are not affected by this concern.

Consult the AERMOD User Guide and AERMOD Implementation Guide for details in applying AERMOD to roadway sources.

Modeling Area Sources

AERMOD can represent rectangular, polygon-shaped, and circular area sources using the AREA, AREAPOLY, or AREACIRC keywords. Sources that may be modeled as area sources may include areas within which emissions occur relatively evenly, such as a single link modeled using MOVES or EMFAC.[10] Evenly-distributed ground-level sources might also be modeled as area sources.

AERMOD requires the following information when modeling an area source:
- The emission rate per unit area (mass per unit area per unit time);
- The release height above the ground;
- The length of the north-south side of the area;
- The length of the east-west side of the area (if the area is not a square);
- The orientation of the rectangular area in degrees relative to north; and
- The initial height (vertical dimension) of the area source plume.

In using a series of area sources to represent emissions of a roadway, the release height and initial vertical dimension of the plume should be calculated as described above for volume sources.

Modeling Point Sources

It may be appropriate to model some emission sources as fixed point sources, such as exhaust fans or stacks on a bus garage or terminal building. If a source is modeled with the POINT keyword in AERMOD, the model requires:
- The emission rate (mass per unit time);
- The release height above the ground;
- The exhaust gas exit temperature;
- The stack gas exit velocity; and,
- The stack inside diameter in meters.

These parameters can often be estimated using the plans and engineering diagrams for ventilation systems.

For projects with emissions on or near rooftops, such as bus terminals or garages, building downwash should also be modeled for the relevant sources. The potential for building downwash should also be addressed for nearby sources whose emissions are on or near rooftops in the project area. Building downwash occurs when air moving over a

---

[10] At present, the area sources in AERMOD do not include AERMOD's "plume meander approach." Consult the latest version of the AERMOD Implementation Guide for the most current information on when volume sources or area sources are most appropriate.

building mixes to the ground on the "downwind" side of the building. AERMOD includes algorithms to model the effects of building downwash on plumes from nearby or adjacent point sources. Consult the AERMOD User Guide for additional detail on how to enter building information.

J.3.4 Placement and sizing of sources within AERMOD

There are several general considerations with regard to placing and sizing sources within AERMOD.

First, volume, area, and point sources should be placed in the locations where emissions are most likely to occur. For example: if buses enter and exit a bus terminal from a single driveway, the driveway should be modeled using one or more discrete volume or area sources in the location of that driveway, rather than spreading the emissions from that driveway across the entire terminal yard.

Second, for emissions from the sides or tops of buildings (as may be found from a bus garage exhaust fan), it may be necessary to use the BPIPPRIME utility in AERMOD to appropriately capture the characteristics of these emissions (such as downwash).

Third, the initial dimensions and other parameters of each source should be as realistic as is feasible. Chapter 3 of the AERMOD User Guide includes recommendations for how to appropriately characterize the shape of area and volume sources.

Finally, if nearby sources are to be included in air quality modeling (see discussion in Section 8 of the guidance), a combination of all these source types may be needed to appropriately represent their emissions within AERMOD. For instance, evenly-distributed ground-level sources might also be modeled as area sources, while a nearby power plant stack might be modeled as a point source.

J.3.5 Timing of emissions in AERMOD

Within AERMOD, emissions that vary across a year should be described with the EMISFACT keyword (see Section 3.3.5 of the AERMOD User Guide). The number of quarters that need to be analyzed may vary based on a particular PM hot-spot analysis. See Section 2.5 of the guidance for more information on when PM emissions need to be evaluated, and Sections 4 and 5 of the guidance on determining the number of MOVES and EMFAC runs.

The *Qflag* parameter under EMISFACT may be used with a secondary keyword to describe different patterns of emission variations throughout a year. Note that AERMOD defines seasons in the following manner: winter (December, January, February), spring (March, April, May), summer (June, July, August), and fall (September, October, November). Emission data obtained from MOVES or EMFAC should be appropriately matched with the relevant time periods in AERMOD. For example, if four MOVES or EMFAC runs are completed (one for each quarter of a year), there are emission estimates corresponding to four months of the year (January, April, July, October) and peak and average periods within each day. In such a circumstance, January runs should be used to represent all AERMOD winter months (December, January, February), April runs for all spring months (March, April, May), July runs for all summer months (June, July, August), and October for all fall months (September, October, November).

If separate weekend emission rates are available, season-specific weekday runs should be used for the Monday-Friday entries; weekend runs would be assigned to the Saturday and Sunday entries. The peak/average runs for each day should be mapped to the AERMOD entry hours corresponding to the relevant time of day from the traffic analysis. *Qflag* can be used to represent emission rates that vary by season, hour of day, and day of the week. Consult the AERMOD User Guide for details.

J.4   INCORPORATING METEOROLOGICAL DATA

This discussion supplements Section 7.5 of the guidance and describes in more detail how to handle meteorological data in AERMOD and CAL3QHCR. Section 7.2.3 of Appendix W to 40 CFR Part 51 provides the basis for determining the urban/rural status of a source. Consult the AERMOD Implementation Guide for instructions on what type of population data should be used in making urban/rural determinations.

J.4.1   Specifying urban or rural sources in AERMOD

As described in Section 7 of the guidance, AERMOD employs nearby population as a surrogate for the magnitude of differential urban-rural heating (i.e., the urban heat island effect). When modeling urban sources in AERMOD, users should use the URBANOPT keyword to enter this data.

When considering urban roughness lengths, users should consult the AERMOD Implementation Guide. Any application of AERMOD that utilizes a value other than 1 meter for the urban roughness length should be considered a non-regulatory application and would require appropriate documentation and justification as an alternate model (see Section 7.3.3 of the guidance).

For urban applications using representative National Weather Service (NWS) meteorological data, consult the AERMOD Implementation Guide. For urban applications using NWS data, the URBANOPT keyword should be selected, regardless of whether the NWS site is located in a nearby rural or urban setting. When using site-specific meteorological data in urban applications, consult the AERMOD Implementation Guide.

J.4.2   Specifying urban or rural sources in CAL3QHCR

CAL3QHCR requires that users specify the run as being rural or urban using the "RU" keyword.[11] Users should make the appropriate entry depending if the source is considered urban or rural as described in Section 7.5.5 of the guidance.

J.5   MODELING COMPLEX TERRAIN

This discussion supplements Section 7.5 of the guidance and describes in more detail how to address complex terrain in AERMOD and CAL3QHCR. In most situations, the project area should be modeled as having flat terrain. Additional detail on how this should be accomplished in each model is found below. However, in some situations a project area may include complex terrain, such that sources and receptors included in the model are found at different heights.

J.5.1   AERMOD

This guidance reflects the AERMOD Implementation Guide as of March 19, 2009. Analysts should consult the most recent AERMOD Implementation Guide for the latest guidance on modeling complex terrain.

For most highway and transit projects, the analyst should apply the non-DFAULT option in AERMOD and assume flat, level terrain. In the AERMOD input file, the FLAT option should be used in the MODELOPT keyword. This recommendation is made to avoid underestimating concentrations in two circumstances likely to occur with the low-elevation, non-buoyant emissions from transportation projects. First, in DFAULT mode, AERMOD will tend to underestimate concentrations from low-level, non-buoyant sources where there is up-sloping terrain with downwind receptors uphill since the DFAULT downwind horizontal plume will pass below the actual receptor elevation. Second, in DFAULT mode, AERMOD will tend to underestimate concentrations when a plume is terrain-following. Therefore, the FLAT option should be selected in most cases.

There may be some cases where significant concentrations result from nearby elevated sources. In these cases, interagency consultation should be used on a case-by-case basis to determine whether to include terrain effects and use the DFAULT option. In those cases, AERMAP should be used to prepare input files for AERMOD; consult the AERMOD and AERMAP user guides and the latest AERMOD Implementation Guide for information on obtaining and processing relevant terrain data.

---

[11] Specifying urban modeling with the "RU" keyword converts stability classes E and F to D.

J.5.2   CAL3QHCR

CAL3QHCR does not handle complex terrain. No action is therefore required.

J.6   RUNNING THE MODEL AND OBTAINING RESULTS

This discussion supplements Section 7.7 of the guidance and describes in more detail how to handle data outputs in AERMOD and CAL3QHCR. AERMOD and CAL3QHCR produce different output file formats, which must be post-processed in different ways to enable calculation of design values as described in Section 9.3 of the guidance. This guidance is applicable regardless of how many quarters are being modeled.

J.6.1   AERMOD output

AERMOD requires that users specify the type and format of output files in the main input file for each run. See Section 3.7 of the AERMOD User Guide for details on the various output options. Output options should be specified to enable the relevant design value calculations required in Section 9.3. Note that many users will have multiple years of meteorological data, so multiple output files may be required (unless the meteorological files have been joined prior to running AERMOD).

For the annual $PM_{2.5}$ design value calculations described in Section 9.3.2, averaging times should be specified that allow calculation of the annual average concentrations at each receptor. For example, when using five years of meteorological data, the PERIOD averaging time could be specified using the CO AVERTIME keyword.

For the 24-hour $PM_{2.5}$ design value calculations described in Section 9.3.3, the DAYTABLE option provides output files with 24-hour concentrations at each receptor for each day processed. Users should flag the quarter and year for each day listed in the DAYTABLE that AERMOD generates. Note users should also specify a 24-hour averaging time with the CO AVERTIME command as well.

Another option for calculating 24-hour $PM_{2.5}$ design values is with a POSTFILE, a file of results at each receptor for each day processed. By specifying a POSTFILE with a 24-hour averaging time, a user can generate a file of daily concentrations for each day of meteorological data. When using this option, users should specify a POSTFILE with a 24-hour averaging time to generate the outputs needed to calculate design values and flag the quarter and year for each day listed in the POSTFILE that AERMOD generates. Note that POSTFILE output files can be very large.

For the 24-hour PM_{10} calculations described in Section 9.3.4, the RECTABLE keyword may be used to obtain the six highest 24-hour concentrations over the entire modeling period. A RECTABLE is a file summarizing the highest concentrations at each receptor over an averaging period (e.g., 24 hours) across a modeling period (e.g., 5 years).

EPA is actively working towards a post-processing tool for AERMOD that will provide the appropriate modeling metrics that may then be combined with background concentrations for comparisons to the PM NAAQS. EPA will announce these new options as they become available on EPA's SCRAM website at: www.epa.gov/scram001/.

J.6.2 CAL3QHCR output

For each year of meteorological data and quarterly emission inputs, CAL3QHCR reports the five highest 24-hour concentrations and the quarterly average concentrations in its output file.

For calculating annual $PM_{2.5}$ design values using CAL3QHCR output, some post-processing is required. CAL3QHCR's output file refers to certain data under the display: "THE HIGHEST ANNUAL AVERAGE CONCENTRATIONS." If four quarters of emission data are separately run in CAL3QHCR, each quarter's outputs listed under "THE HIGHEST ANNUAL AVERAGE CONCENTRATIONS" are actually quarterly-average concentrations. As described in Section 7, per year of meteorological data, CAL3QHCR should be run for as many quarters as analyzed using MOVES and EMFAC, as CAL3QHCR accepts only a single quarter's emission factors per input file.

Calculating 24-hour $PM_{2.5}$ design values under a first or second tier analysis is described in Section 9.3.3. To get annual average modeled concentrations for a first tier analysis (Step 1), the highest 24-hour concentrations in each quarter and year of meteorological data should be identified. Within each year of meteorological data, the highest 24-hour concentration at each receptor should be identified. For a first tier analysis, at each receptor, the highest concentrations from each year of meteorological data should be averaged together. Under a second tier analysis, at each receptor, the highest modeled concentration in each quarter, from each year of meteorological data, should be averaged together. These average highest 24-hour concentrations in each quarter, across multiple years of meteorological data, are used in second tier $PM_{2.5}$ design value calculations.

In calculating 24-hour PM_{10} design values, it is necessary to estimate the sixth-highest concentration in each year if using five years of meteorological data. For each period of meteorological data, CAL3QHCR outputs the five highest 24-hour concentrations. To estimate the sixth-highest concentration at a receptor, the five highest 24-hour concentrations from each quarter and year of meteorological data should be arrayed together and ranked. From all quarters and years of meteorological data, the sixth-highest concentration should be identified. This concentration, at each receptor, is used in calculations of the PM_{10} design value described in Section 9.3.4.

Appendix K:
Examples of Design Value Calculations for PM Hot-spot Analyses

K.1 INTRODUCTION

This appendix supplements Section 9's discussion of calculating and applying design values for PM hot-spot analyses. While this guidance can apply to any PM NAAQS, this appendix provides examples of how to calculate design values for the PM NAAQS in effect at the time the guidance was issued (the 1997 annual $PM_{2.5}$ NAAQS, the 2006 and 1997 24-hour $PM_{2.5}$ NAAQS, and the 1987 24-hour PM_{10} NAAQS). The design values in this appendix are calculated using the steps described in Section 9.3. Readers should reference the appropriate sections of the guidance as needed for more detail on how to complete each step of these analyses.

These illustrative example calculations demonstrate the basic procedures described in the guidance and therefore are simplified in the number of receptors considered and other details that would occur in an actual PM hot-spot analysis. Where users would have to repeat steps for additional receptors, it is noted. These examples are organized according to the build/no-build analysis steps that are described in Sections 2 and 9 of this guidance.

The final part of this appendix provides mathematical formulas that describe the design value calculations discussed in Section 9 and this appendix.

K.2 PROJECT DESCRIPTION AND CONTEXT FOR ALL EXAMPLES

For the following examples, a PM hot-spot analysis is being done for an expansion of an existing highway with a significant increase in the number of diesel vehicles (40 CFR 93.123(b)(1)(i)). The highway expansion will serve an expanded freight terminal. The traffic at the terminal will increase as a result of the expanded highway project's increase in truck traffic, and therefore the freight terminal is projected to have higher emissions under the build scenario than under the no-build scenario. The freight terminal is not part of the project; however, it is a nearby source that will be included in the air quality modeling, as described further below.

The air quality monitor selected to represent background concentrations from other sources is a Federal Equivalent Method (FEM) monitor that is 300 meters upwind of the project. The monitor is on a 1-in-3 day sampling schedule. In this example, the three most recent years of monitoring data are from 2008, 2009, and 2010. Since 2008 is a leap year (366 days), for this example, there are 122 monitored values in that year and 121 values for both 2009 and 2010 (365 days each).[1]

---

[1] Note that the number of air quality monitoring measurements may vary by year. For example, with 1-in-3 measurements, there could be 122 or 121 measurements in a year with 365 days. Or, there may be fewer

However, through interagency consultation, it is determined that the freight terminal's emissions are not already captured by this air quality monitor. AERMOD has been selected as the air quality model to estimate PM concentrations produced by the project (the highway expansion) and the nearby source (the freight terminal).[2] There are five years of representative off-site meteorological data being used in this analysis.

As discussed in Section 2.4, a project sponsor could consider mitigation and control measures at any point in the process. However, since the purpose of these examples is to show the design value calculations, in this appendix such measures are not considered until after the calculations are done.

K.3 EXAMPLE: ANNUAL PM$_{2.5}$ NAAQS

K.3.1 General

This example illustrates the approach to calculating design values for comparison to the annual PM$_{2.5}$ NAAQS, as described in Section 9.3.2. The annual PM$_{2.5}$ design value is the average of three consecutive years' annual averages. The design value for comparison is rounded to the nearest tenth of a $\mu g/m^3$ (nearest 0.1 $\mu g/m^3$). For example, 15.049 rounds to 15.0, and 15.050 rounds to 15.1.[3]

Each year's annual average concentrations include contributions from the project, any nearby sources modeled, and background concentrations. For air quality monitoring purposes, the annual PM$_{2.5}$ NAAQS is met when the three-year average concentration is less than or equal to the current annual PM$_{2.5}$ NAAQS (i.e., 15.0 $\mu g/m^3$):

Annual PM$_{2.5}$ design value = ([Y1] average + [Y2] average + [Y3] average) ÷ 3

> Where:
> [Y1] = Average annual PM$_{2.5}$ concentration for the <u>first</u> year of air quality monitoring data
> [Y2] = Average annual PM$_{2.5}$ concentration for the <u>second</u> year of air quality monitoring data
> [Y3] = Average annual PM$_{2.5}$ concentration for the <u>third</u> year of air quality monitoring data

---

actual monitored values if sampling was not conducted on some scheduled days or the measured value was invalidated due to quality assurance concerns. The actual number of samples with valid data should be used.

[2] EPA notes that CAL3QHCR could not be used in this particular PM hot-spot analysis, since air quality modeling included the project and a nearby source. See Section 7.3 of the guidance for further information.

[3] A sufficient number of decimal places (3-4) should be retained during intermediate calculations for design values, so that there is no possibility of intermediate rounding or truncation affecting the final result. Rounding to the tenths place should only occur during final design value calculations, pursuant to Appendix N to 40 CFR Part 50.

For this example, the project described in Appendix K.2 is located in an annual $PM_{2.5}$ NAAQS nonattainment area. This example illustrates how an annual $PM_{2.5}$ design value could be calculated at the same receptor in the build and no-build scenarios, based on air quality modeling results and air quality monitoring data. In an actual PM hot-spot analysis, design values would be calculated at additional receptors, as described further in Section 9.3.2.

K.3.2   Build scenario

For the build scenario, the $PM_{2.5}$ impacts from the project and from the nearby source are estimated with AERMOD at all receptors.[4]

Steps 1-2. Because AERMOD is used for this project, Step 1 is skipped. The receptor with the highest average annual concentration, using five years of meteorological data, is identified directly from the AERMOD output. This receptor's average annual concentration is 3.603 $\mu g/m^3$.

Step 3. Based on the three years of measurements at the background air quality monitor, the average monitored background concentrations in each quarter is determined. Then, for each year of background data, the four quarters are averaged to get an average annual background concentration (last column of Exhibit K-1). These three average annual background concentrations are averaged, and the resulting value is 11.582 $\mu g/m^3$, as shown in Exhibit K-1:

Exhibit K-1. Background Concentrations

Background Concentrations	Q1	Q2	Q3	Q4	Average Annual
2008	13.013	17.037	8.795	8.145	11.748
2009	14.214	14.872	7.912	7.639	11.159
2010	11.890	16.752	9.421	9.287	11.838
3-year average:					11.582

Step 4. The 3-year average annual background concentration (from Step 3) is added to the average annual modeled concentration from the project and nearby source (from Step 2):

$$11.582 + 3.603 = 15.185$$

Step 5. Rounding to the nearest 0.1 $\mu g/m^3$ produces a design value of 15.2 $\mu g/m^3$.

---

[4] As noted above, there is a single nearby source that is projected to have higher emissions under the build scenario than the no-build scenario as a result of the project and its impacts are not expected to be captured by the monitor chosen to provide background concentrations. Therefore, emissions from the project and this nearby source are both included in the AERMOD output.

In this example, the concentration at the highest receptor is estimated to exceed the current annual $PM_{2.5}$ NAAQS of 15.0 $\mu g/m^3$.

Steps 6-8: Since the design value in Step 5 is greater than the NAAQS, design value calculations are then completed for all receptors in the build scenario, and receptors with design values above the NAAQS are identified. After this is done, the no-build scenario is modeled for comparison.

K.3.3 No-build scenario

The no-build scenario (i.e., the existing highway and freight terminal without the proposed highway and freight terminal expansion), is modeled at all of the receptors in the build scenario, but design values are only calculated in the no-build scenario at receptors where the design value for the build scenario is above the annual $PM_{2.5}$ NAAQS (from Steps 6-8 above).

Step 9. For this example, the receptor with the highest average annual concentration in the build scenario is used to illustrate the no-build scenario design value calculation. The average annual concentration modeled at this receptor in the no-build scenario is 3.521 $\mu g/m^3$.

Step 10. The background concentrations from the representative monitor are unchanged from the build scenario, so the average annual modeled concentration of 3.521 is added to the 3-year average annual background concentrations of 11.528 $\mu g/m^3$ from Step 3:

$$11.582 + 3.521 = 15.103$$

Step 11. Rounding to the nearest 0.1 $\mu g/m^3$ produces a design value of 15.1 $\mu g/m^3$.

In this example, the design value at the receptor in the build scenario (15.2 $\mu g/m^3$) is greater than the design value at the same receptor in the no-build scenario (15.1 $\mu g/m^3$).[5] In an actual PM hot-spot analysis, design values would also be compared between build and no-build scenarios at all receptors in the build scenario that exceeded the annual $PM_{2.5}$ NAAQS. The interagency consultation process would then be used to discuss next steps, e.g., appropriateness of receptors. Refer to Sections 9.2 and 9.4 for additional details.

If it is determined that conformity requirements are not met at all appropriate receptors, the project sponsor should then consider additional mitigation or control measures, as discussed in Section 10. After measures are selected, a new build scenario that includes the controls should be modeled and new design values calculated. Design values for the no-build scenario shown above would not need to be recalculated since the no-build scenario would not change.

---

[5] Values are compared after rounding. As long as the build design value is no greater than the no-build design value after rounding, the project would meet conformity requirements at a given receptor, even if the pre-rounding build design value is greater than the pre-rounding no-build design value.

K.4 EXAMPLE: 24-HOUR PM₂.₅ NAAQS

K.4.1 General

This example illustrates the two-tiered approach to calculating design values for comparison with the 24-hour $PM_{2.5}$ NAAQS, as described in Section 9.3.3. The 24-hour design value is the average of three consecutive years' 98th percentile $PM_{2.5}$ concentration of 24-hour values for each of those years. For air quality monitoring purposes, the NAAQS is met when that three-year average concentration is less than or equal to the currently applicable 24-hour $PM_{2.5}$ NAAQS for a given area's nonattainment designation (35 $\mu g/m^3$ for nonattainment areas for the 2006 $PM_{2.5}$ NAAQS and 65 $\mu g/m^3$ for nonattainment areas for the 1997 $PM_{2.5}$ NAAQS).[6] The design value for comparison to any 24-hour $PM_{2.5}$ NAAQS is rounded to the nearest 1 $\mu g/m^3$ (i.e., decimals 0.5 and greater are rounded up to the nearest whole number, and any decimal lower than 0.5 is rounded down to the nearest whole number). For example, 35.499 rounds to 35 $\mu g/m^3$, while 35.500 rounds to 36.[7]

For this example, the project described in Appendix K.2 is located in a nonattainment area for the 2006 24-hour $PM_{2.5}$ NAAQS. This example presents first tier and second tier build scenario results for a single receptor to illustrate how the calculations should be made based on air quality modeling results and air quality monitoring data. It also shows second tier no-build scenario results for this same receptor. In an actual PM hot-spot analysis, design values would be calculated at additional receptors, as described further in Section 9.3.3.

As explained in Section 9.3.3, project sponsors can start with either a first or second tier analysis. This example begins with a first tier analysis. However, it would also be acceptable to begin with the second tier analysis and skip the first tier altogether.

K.4.2 Build scenario

$PM_{2.5}$ contributions from the project and the nearby source are estimated together with AERMOD in each of four quarters using meteorological data from five consecutive years, using a 24-hour averaging time. As discussed in Appendix K.2 above, the one nearby source (the freight terminal) was included in air quality modeling.

---

[6] There are only two $PM_{2.5}$ areas where conformity currently applies for both the 1997 and 2006 24-hour NAAQS. While both 24-hour NAAQS must be considered in these areas, in practice if the more stringent 2006 24-hour $PM_{2.5}$ NAAQS is met, then the 1997 24-hour $PM_{2.5}$ NAAQS is met as well.
[7] A sufficient number of decimal places (3-4) should be retained during intermediate calculations for design values, so that there is no possibility of intermediate rounding or truncation affecting the final result. Rounding should only occur during final design value calculations, pursuant to Appendix N to 40 CFR Part 50.

First Tier Analysis

Under a first tier analysis, the average highest modeled 24-hour concentrations at a given receptor are added to the average 98th percentile 24-hour background concentrations, regardless of the quarter in which they occur. The average highest modeled 24-hour concentrations are produced by AERMOD, using five years of meteorological data in one run.

Step 1. The receptor with the highest average modeled 24-hour concentration is identified. This was obtained directly from the AERMOD output.[8] For this example, the data from this receptor is shown in Exhibit K-2. Exhibit K-2 shows the highest 24-hour concentration for each year of meteorological data used, regardless of the quarter in which they were modeled. The average concentration of these outcomes, 6.710 μg/m^3 (highlighted in Exhibit K-2), is the highest, compared to the averages at all of the other receptors.

Exhibit K-2. Modeled PM$_{2.5}$ Concentrations from Project and Nearby Source

Year	Highest PM$_{2.5}$ Concentration
Met Year 1	6.413
Met Year 2	5.846
Met Year 3	6.671
Met Year 4	7.951
Met Year 5	6.667
Average	6.710

Step 2. The average 98th percentile 24-hour background concentration for a first tier analysis is calculated using the 98th percentile 24-hour concentrations of the three most recent years of monitoring data from the representative air quality monitor selected (see Appendix K.2). Since the background monitor is on a 1-in-3 day sampling schedule, it made either 122 or 121 measurements per year during 2008 - 2010. According to Exhibit 9-5, with this number of monitored values per year, the 98th percentile is the third highest concentration. Exhibit K-3 depicts the top eight monitored concentrations (in μg/m^3) of the monitor throughout the years employed for estimating background concentrations.

---

[8] If CAL3QHCR were being used, some additional processing of model output would be needed. Refer to Section 9.3.3.

Exhibit K-3. Top Eight Monitored Concentrations in Years 2008 – 2010

Rank	2008	2009	2010
1	34.123	33.537	35.417
2	31.749	32.405	31.579
3	31.443	31.126	31.173
4	30.809	30.819	31.095
5	30.219	30.487	30.425
6	30.134	29.998	30.329
7	30.099	29.872	30.193
8	28.481	28.937	28.751

The third-ranked concentration of each year (highlighted in Exhibit K-3) is the 98th percentile value. These are averaged:

$$(31.443 + 31.126 + 31.173) \div 3 = 31.247 \ \mu g/m^3.$$

Step 3. Then, the highest average 24-hour modeled concentration for this receptor (from Step 1) is added to the average 98th percentile 24-hour background concentration (from Step 2):

$$6.710 + 31.247 = 37.957 \ \mu g/m^3.$$

Rounding to the nearest whole number results in a 24-hour $PM_{2.5}$ design value of 38 $\mu g/m^3$.

Because this concentration is greater than the 2006 24-hour $PM_{2.5}$ NAAQS (35 $\mu g/m^3$), this first tier analysis does not demonstrate that conformity is met. As described in Section 9.3.3, the project sponsor has two options:
- Repeat the first tier analysis for the no-build scenario at all receptors that exceeded the NAAQS in the build scenario. If the calculated design value for the build scenario is less than or equal to the design value for the no-build scenario at all of these receptors, then the project conforms;[9] or
- Conduct a second tier analysis.

In this example, the next step chosen is to conduct a second tier analysis.

Second Tier Analysis

In a second tier analysis, the highest modeled concentrations are not added to the 98th percentile background concentrations on a yearly basis. Instead, a second tier analysis uses the average of the highest modeled 24-hour concentration within each quarter of each year of meteorological data. Impacts from the project, nearby sources, and other background concentrations are calculated on a quarterly basis before determining the 98th

---

[9] In certain cases, project sponsors can also decide to calculate the design values for all receptors in the build and no-build scenarios and use the interagency consultation process to determine whether a "new" violation has been relocated (see Section 9.2).

percentile concentration resulting from these inputs. The steps presented below follow the steps described in Section 9.3.3.

Step 1. The first step is to count the number of measurements for each year of monitoring data used for background concentrations. As described in Appendix K.2 and in Step 2 of the first tier analysis above, there are 122 monitored values during 2008, 121 values during 2009, and 121 values during 2010.

Step 2. For each year of monitoring data, the eight highest 24-hour background concentrations in each quarter are determined. The eight highest concentrations in each quarter of 2008, 2009, and 2010 are shown in Exhibit K-4.

Exhibit K-4. Eight Highest 24-hour Background Concentrations By Quarter for Each Year

Year	Rank of Background Concentration	Q1	Q2	Q3	Q4
2008	1	27.611	31.749	34.123	30.099
	2	25.974	30.219	31.443	28.096
	3	25.760	30.134	30.809	26.990
	4	25.493	28.368	28.481	25.649
	5	25.099	27.319	27.372	25.526
	6	24.902	25.788	25.748	25.509
	7	24.780	25.564	25.288	25.207
	8	23.287	24.794	24.631	24.525
2009	1	26.962	32.405	33.537	31.126
	2	24.820	30.487	30.819	28.553
	3	24.330	28.937	29.998	25.920
	4	23.768	27.035	29.872	25.856
	5	23.685	25.880	25.596	25.565
	6	23.287	25.867	25.148	24.746
	7	23.226	25.254	24.744	24.147
	8	22.698	24.268	24.267	23.142
2010	1	27.493	31.579	35.417	30.425
	2	24.637	31.173	31.095	26.927
	3	24.637	30.193	30.329	26.263
	4	24.392	27.994	28.751	25.684
	5	24.050	25.439	26.084	25.170
	6	23.413	24.253	24.890	24.254
	7	22.453	23.006	24.749	23.425
	8	22.061	21.790	22.538	22.891

Step 3. The highest modeled 24-hour concentrations in each quarter are identified at each receptor. Exhibit K-5 presents the highest 24-hour concentrations within each quarter at one receptor (for each of the five years of meteorological data used in air quality modeling) as well as the average of these quarterly concentrations. This step would be repeated for each receptor in an actual PM hot-spot analysis.

Exhibit K-5. Highest Modeled 24-hour Concentrations Within Each Quarter (Build Scenario)

	Q1	Q2	Q3	Q4
Met Year 1	6.413	3.332	6.201	6.193
Met Year 2	3.229	3.481	5.846	4.521
Met Year 3	6.671	3.330	5.696	6.554
Met Year 4	7.095	3.584	7.722	7.951
Met Year 5	6.664	4.193	4.916	6.667
Average	6.014	3.584	6.076	6.377

The average highest concentrations on a quarterly basis (i.e., the values highlighted in Exhibit K-5) constitute the contributions of the project and nearby source to the projected 24-hour $PM_{2.5}$ design value, and are used in subsequent calculations.

Step 4. For each receptor, the highest modeled 24-hour concentration in each quarter (from Step 3) is added to each of the eight highest monitored concentrations for the same quarter for each year of monitoring data (from Step 2). To obtain this result, the average highest modeled concentration for each quarter, found in the last row of Exhibit K-5, is added to each of the eight highest background concentrations in each quarter in Exhibit K-4. The results are shown in Exhibit K-6.

Exhibit K-6. Sum of Modeled and Monitored Concentrations (Build Scenario)

Year	Rank of Background Concentration	Q1	Q2	Q3	Q4
2008	1	33.625	35.333	40.200	36.476
	2	31.989	33.803	37.520	34.474
	3	31.774	33.718	36.886	33.368
	4	31.507	31.952	34.557	32.026
	5	31.113	30.903	33.448	31.903
	6	30.916	29.372	31.824	31.886
	7	30.794	29.148	31.365	31.584
	8	29.301	28.378	30.707	30.902
2009	1	32.976	35.989	39.613	37.503
	2	30.835	34.071	36.895	34.931
	3	30.344	32.521	36.074	32.297
	4	29.782	30.619	35.948	32.233
	5	29.700	29.464	31.672	31.942
	6	29.301	29.451	31.225	31.124
	7	29.240	28.838	30.820	30.524
	8	28.712	27.852	30.343	29.520
2010	1	33.507	35.163	41.493	36.802
	2	30.651	34.757	37.172	33.304
	3	30.651	33.777	36.405	32.640
	4	30.406	31.578	34.827	32.062
	5	30.064	29.022	32.160	31.547
	6	29.428	27.837	30.966	30.631
	7	28.468	26.590	30.825	29.803
	8	28.075	25.374	28.614	29.269

Step 5. The 32 values from each year in Exhibit K-6 are then ranked from highest to lowest, regardless of the quarter from which each value comes. This step is shown in Exhibit K-7. Note that only the top eight values are shown for each year instead of the entire set of 32. Exhibit K-7 also displays the quarter from which each concentration comes and the value's rank within its quarter.

Exhibit K-7. Eight Highest Concentrations in Each Year, Ranked from Highest to Lowest (Build Scenario)

Year	μg/m³	Yearly Rank	Quarter	Quarterly Rank
2008	40.200	1	Q3	1
	37.520	2	Q3	2
	36.886	3	Q3	3
	36.476	4	Q4	1
	35.333	5	Q2	1
	34.557	6	Q3	4
	34.474	7	Q4	2
	33.803	8	Q2	2
2009	39.613	1	Q3	1
	37.503	2	Q4	1
	36.895	3	Q3	2
	36.074	4	Q3	3
	35.989	5	Q2	1
	35.948	6	Q3	4
	34.931	7	Q4	2
	34.071	8	Q2	2
2010	41.493	1	Q3	1
	37.172	2	Q3	2
	36.802	3	Q4	1
	36.405	4	Q3	3
	35.163	5	Q2	1
	34.827	6	Q3	4
	34.757	7	Q2	2
	33.777	8	Q2	3

Steps 6-7. The value that represents the 98th percentile 24-hour concentration is determined, based on the number of background concentration values there are. As described in Step 1, there are 122 monitored values for the year 2008 and 121 values for both 2009 and 2010. According to Exhibit 9-7 in Section 9.3.3, for a year with 101-150 samples per year, the 98th percentile is the 3rd highest concentration for that year. Therefore, for this example, the 3rd highest 24-hour concentration of each year, highlighted in Exhibit K-7, represents the 98th percentile value for that year.

Step 8. At each receptor, the average of the three 24-hour 98th percentile concentrations is calculated. For the receptor in this example, the average is:
    (36.886 + 36.895 + 36.802) ÷ 3 = 36.861

Step 9. The average for the receptor in this example from Step 8 (36.861 $\mu g/m^3$) is then rounded to the nearest whole number (37 $\mu g/m^3$) and compared to the 2006 24-hour $PM_{2.5}$ NAAQS (35 $\mu g/m^3$).

The design value at the receptor in this example is higher than the relevant 24-hour $PM_{2.5}$ NAAQS. In an actual $PM_{2.5}$ hot-spot analysis, the design value calculations need to be repeated for all receptors, and compared to the NAAQS. Since one (and possibly more) receptors have design values greater than the 24-hour $PM_{2.5}$ NAAQS, the project will only conform if the design value in the build scenario is less than or equal to the design value in the no-build scenario for all receptors that exceeded the NAAQS in the build scenario. Therefore, the no-build scenario needs to be modeled for comparison, as described further below. Because the build scenario was modeled with a second tier analysis, the no-build scenario must also be modeled with a second tier analysis.

K.4.3 No-build scenario

The no-build scenario is described in Section 9.3.3 as Step 10:
- Step 10. Using modeling results for the no-build scenario, repeat Steps 3 through 9 for all receptors with a design value that exceeded the $PM_{2.5}$ NAAQS in the build scenario. The result will be a 24-hour $PM_{2.5}$ design value at such receptors for the no-build scenario.

For this part of the example, air quality modeling is completed for the no-build scenario for the same receptor as the build scenario. Steps 1 and 2 for the build scenario do not need to be repeated, since the background concentrations in the no-build scenario are identical to those in the build scenario. Exhibit K-4, which shows the eight highest monitored concentrations in each quarter over three years, therefore can also be used for the no-build scenario.

Step 3. For the same receptor examined above in the build scenario, the highest modeled 24-hour concentrations for the no-build scenario are calculated for each quarter, using each year of meteorological data used for air quality modeling. Exhibit K-8 provides these concentrations, as well as the quarterly averages (highlighted).

Exhibit K-8. Highest Modeled 24-hour Concentrations Within Each Quarter (No-Build Scenario)

	Q1	Q2	Q3	Q4
Met Year 1	6.757	3.383	6.725	6.269
Met Year 2	3.402	3.535	6.340	4.577
Met Year 3	7.029	3.381	6.177	6.635
Met Year 4	7.476	3.639	8.374	8.048
Met Year 5	7.022	4.258	5.331	6.748
Average	6.337	3.639	6.589	6.455

Step 4. The highest modeled 24-hour concentration in each quarter (i.e., the values in the last row of Exhibit K-8) are added to each of the eight highest concentrations for the same quarter for each year of monitoring data (found in Exhibit K-4), and the resulting values are shown in Exhibit K-9.

Exhibit K-9. Sum of Modeled and Monitored Concentrations (No-Build Scenario)

Year	Rank of Background Concentration	Q1	Q2	Q3	Q4
2008	1	33.948	35.389	40.713	36.555
	2	32.312	33.858	38.033	34.552
	3	32.097	33.774	37.399	33.446
	4	31.830	32.007	35.070	32.104
	5	31.436	30.959	33.961	31.981
	6	31.239	29.428	32.337	31.964
	7	31.117	29.204	31.878	31.662
	8	29.624	28.433	31.220	30.980
2009	1	33.299	36.044	40.127	37.581
	2	31.158	34.126	37.408	35.009
	3	30.667	32.576	36.587	32.375
	4	30.105	30.674	36.461	32.311
	5	30.023	29.520	32.185	32.020
	6	29.624	29.506	31.738	31.202
	7	29.563	28.894	31.333	30.602
	8	29.035	27.907	30.856	29.598
2010	1	33.830	35.218	42.007	36.880
	2	30.974	34.812	37.685	33.382
	3	30.974	33.832	36.918	32.719
	4	30.729	31.633	35.340	32.140
	5	30.387	29.078	32.674	31.625
	6	29.751	27.893	31.479	30.709
	7	28.791	26.645	31.338	29.881
	8	28.398	25.429	29.127	29.347

Step 5. The 32 values from each year in Exhibit K-9 are ranked from highest to lowest, regardless of the quarter from which each value comes. This step is shown in Exhibit K-10. Note that only the top eight values are shown for each year instead of the entire set of 32.

Exhibit K-10. Eight Highest Concentrations in Each Year, Ranked from Highest to Lowest (No-Build Scenario)

Year	µg/m³	Yearly Rank	Quarter	Quarterly Rank
2008	40.713	1	Q3	1
	38.033	2	Q3	2
	37.399	3	Q3	3
	36.555	4	Q4	1
	35.389	5	Q2	1
	35.070	6	Q3	4
	34.552	7	Q4	2
	33.961	8	Q3	5
2009	40.127	1	Q3	1
	37.581	2	Q4	1
	37.408	3	Q3	2
	36.587	4	Q3	3
	36.461	5	Q3	4
	36.044	6	Q2	1
	35.009	7	Q4	2
	34.126	8	Q2	2
2010	42.007	1	Q3	7
	37.685	2	Q1	3
	36.918	3	Q1	2
	36.880	4	Q4	8
	35.340	5	Q4	6
	35.218	6	Q1	1
	34.812	7	Q4	2
	33.832	8	Q4	3

Steps 6-7. Based on the number of background measurements available per year in this example (122 for 2008 and 121 for both 2009 and 2010, as discussed in the analysis of the build scenario), Exhibit 9-7 in Section 9.3.3 indicates that the 3rd highest 24-hour concentration in each year represents the 98th percentile concentration for that year. The third highest concentrations are highlighted in Exhibit K-10.

Step 8. For this receptor, the average of the Rank 3 concentrations in 2008, 2009, and 2010 is calculated:

$$(37.399 + 37.408 + 36.918) \div 3 = 37.242$$

Step 9. The average for the receptor in this example from Step 8 (37.242 µg/m³) is rounded to the nearest whole µg/m³ (37 µg/m³).

In this example, the design value at this receptor for both the build and no-build scenarios is 37 µg/m³, which is greater than the 2006 24-hour NAAQS (35 mg/m³). However, the build scenario's design value is equal to the design value in the no-build scenario.[10] For the project to conform, the build design values must be less than or equal to the no-build value for all the receptors that exceeded the NAAQS in the build scenario. Assuming that this is the case at all other receptors, the proposed project in this example would therefore demonstrate conformity.

K.5 EXAMPLE: 24-HOUR PM_{10} NAAQS

K.5.1 General

This example illustrates calculating design values for comparison with the 24-hour PM_{10} NAAQS, as described in Section 9.3.4. The 24-hour PM_{10} design value is based on the expected number of 24-hour exceedances of 150 µg/m³, averaged over three consecutive years. For air quality monitoring purposes, the NAAQS is met when the number of exceedances is less than or equal to 1.0. The 24-hour PM_{10} design value is rounded to the nearest 10 µg/m³. For example, 155.511 rounds to 160, and 154.999 rounds to 150.[11]

The 24-hour PM_{10} design value is calculated at each air quality modeling receptor by directly adding the sixth-highest modeled 24-hour concentration (if using five years of meteorological data) to the highest 24-hour background concentration (from three years of monitored data).

For this example, the project described in Appendix K.2 is located in a nonattainment area for the 24-hour PM_{10} NAAQS. This example presents build scenario results for a single receptor to illustrate how the calculations should be made based on air quality modeling results and air quality monitoring data.

---

[10] Values are compared after rounding. As long as the build design value is no greater than the no-build design value after rounding, the project would meet conformity requirements at a given receptor, even if the pre-rounding build design value is greater than the pre-rounding no-build design value.

[11] A sufficient number of decimal places (3-4) in modeling results should be retained during intermediate calculations for design values, so that there is no possibility of intermediate rounding or truncation affecting the final result. Rounding to the nearest 10 ug/m³ should only occur during final design value calculations, pursuant to Appendix K to 40 CFR Part 50. Monitoring values typically are reported with only one decimal place.

K.5.2   Build Scenario

Step 1. From the air quality modeling results from the build scenario, the sixth-highest 24-hour concentration is identified at each receptor. These sixth-highest concentrations are the sixth highest that are modeled at each receptor, regardless of year of meteorological data used.[12] AERMOD was configured to produce these values.

Step 2. The sixth-highest modeled concentrations (i.e., the concentrations at Rank 6) are compared across receptors, and the receptor with the highest value at Rank 6 is identified. For this example, the highest sixth-highest 24-hour concentration at any receptor is 15.218 $\mu g/m^3$. (That is, at all other receptors, the sixth-highest concentration is less than 15.218 $\mu g/m^3$.) Exhibit K-11 shows the six highest 24-hour concentrations at this receptor.

Exhibit K-11. Receptor with the Highest Sixth-Highest 24-Hour Concentration (Build Scenario)

Rank	Highest 24-Hour Concentrations
1	17.012
2	16.709
3	15.880
4	15.491
5	15.400
6	15.218

Step 3. The highest 24-hour background concentration from the three most recent years of monitoring data (2008, 2009, and 2010) is identified. In this example, the highest 24-hour background concentration from these three years is 86.251 $\mu g/m^3$.

Step 4. The sixth-highest 24-hour modeled concentration of 15.218 $\mu g/m^3$ from the highest receptor (from Step 2) is added to the highest 24-hour background concentration of 86.251 $\mu g/m^3$ (from Step 3):
$$15.218 + 86.251 = 101.469$$

Step 5. This sum is rounded to the nearest 10 $\mu g/m^3$, which results in a design value of 100 $\mu g/m^3$.

This result is then compared to the 24-hour PM$_{10}$ NAAQS. In this case, the concentration calculated at all receptors is less than the 24-hour PM$_{10}$ NAAQS of 150 $\mu g/m^3$, therefore

---

[12] The six highest concentrations could occur anytime during the five years of meteorological data. They may be clustered in one or two years, or they may be spread out over several, or even all five, years of the meteorological data.

the analysis shows that the project conforms. However, if the design value for this receptor had been greater than 150 $\mu g/m^3$, the remainder of the steps in Section 9.3.4 would be completed. That is, build scenario design values for each receptor would be calculated (Steps 6-7 in Section 9.3.4) and, for all those that exceed the NAAQS, the no-build design values would also be calculated (Steps 8-10 in Section 9.3.4). The build and no-build design values would then be compared.[13]

K.6   MATHEMATICAL FORMULAS FOR DESIGN VALUE CALCULATIONS

K.6.1   Introduction

This part of the appendix includes mathematical formulas to represent the calculations described narratively in Section 9.3. This information is intended to supplement Section 9, which may be helpful for certain users.

Appendix K.6 relies on conventions of mathematical and logical notation that are described after the formulas are presented. Several symbols are used that may be useful to review prior to reading the individual formulas.

<u>Notation symbols</u>

- $\bar{x}$ - a single bar over variable x represents a single arithmetic mean of that variable
- $\bar{\bar{x}}$ - double bars over variable x represents an "average of averages"
- $\hat{x}$ - a "hat" over variable x represents the arithmetic of multiple high concentration values from different years, either from monitoring data or from modeling results

<u>Logical symbols</u>

- $\forall x$ - an upside down A before variable x means "for all" values of x
- $\in x$ - an " $\in$ " before variable x means "in x"
- $\forall x \in y$ - means "for all x in y"

The following information present equations for calculating design values for the PM$_{2.5}$ annual NAAQS, 24-hour PM$_{2.5}$ NAAQS, and 24-hour PM$_{10}$ NAAQS. The equations are organized into the sets that are referenced in Section 9.3.

---

[13] Values are compared after rounding. As long as the build design value is no greater than the no-build design value after rounding, the project would meet conformity requirements at a given receptor, even if the pre-rounding build design value is greater than the pre-rounding no-build design value.

K.6.2 Equation Set 1: Annual PM$_{2.5}$ design value

Formulas

$$\overline{\overline{c}}_i = \overline{\overline{b}}_i + \overline{\overline{p}}_i$$

$$\overline{\overline{b}}_i = \sum_{m=1}^{3} \frac{\overline{b}_{im}}{3}$$

$$\overline{b}_{im} = \sum_{j=1}^{4} \frac{\overline{b}_{ijm}}{4}$$

$$\overline{\overline{p}}_i = \sum_{k=1}^{l} \frac{\overline{p}_{ik}}{l}$$

When using CAL3QHCR, $\overline{p}_{ik} = \sum_{j=1}^{4} \dfrac{\overline{p}_{ijk}}{4}$

Definitions

$\overline{\overline{b}}_i$ = average of three consecutive years' average annual background concentrations at receptor i

$\overline{b}_{im}$ = quarterly-weighted average annual background concentrations at receptor i during monitoring year m

$\overline{b}_{ijm}$ = quarterly average background concentration at receptor i, during quarter j in monitoring year m

$\overline{\overline{c}}_i$ = annual PM$_{2.5}$ design value at receptor i

i = receptor

j = quarter

k = year of meteorological data

l = length in years of meteorological data record

m = year of background monitoring data

$\overline{p}_{ik}$ = average modeled quarterly average concentrations at receptor i for meteorological year k. When using AERMOD, it is presumed that AERMOD's input file is used to specify this averaging time. When using CAL3QHCR with a single quarter of meteorological data, $\overline{p}_{ik}$ must be calculated using each $\overline{p}_{ijk}$ for each quarter of meteorological year k.

$\overline{p}_{ijk}$ = quarterly average concentration at receptor i for quarter j, in meteorological data year k. This variable is the product of CAL3QHCR when run with a single quarter of meteorological data. $\overline{p}_{ik}$ can be calculated directly using AERMOD without explicitly calculating $\overline{p}_{ijk}$.

K.6.3 *Equation Set 2: 24-Hour PM$_{2.5}$ design value (First Tier Analysis)*

Formulas

$$\hat{c}_i = \hat{b}_i + \hat{p}_i$$

$$b_{im} = \forall b_{ijm} \in m$$

$$\hat{b}_i = \sum_{m=1}^{3} \frac{b_{im \bullet r_m}}{3}$$

$$\hat{p}_i = \sum_{k=1}^{l} \frac{\max_k[\max_{jk}(p_{ijk})]}{l} \text{ (when using CAL3QHCR), which compresses to:}$$

$$\hat{p}_i = \sum_{k=1}^{l} \frac{\max_k(p_{ik})}{l} \text{ (when using AERMOD with maximum concentration by year)}$$

Definitions

$\hat{b}_i$ = the average of 98th percentile 24-hour concentrations from three consecutive years of monitoring data

b_{ijm} = daily 24-hour background concentration at receptor i, during quarter j in monitoring year m

$b_{im} = \forall b_{ijm} \in m$ = All 24-hour background concentration measurements in year m

$b_{im \bullet r_m}$ = The 24-hour period within year m whose concentration rank among all 24-hour measurements in year m is r_m (this represents the 98th percentile of 24-hour background concentrations within one year.)

$\hat{c}_i$ = 24-hour PM$_{2.5}$ design value at receptor i

i = receptor

j = quarter

k = year of meteorological data

l = length in years of meteorological data record

m = year of background monitoring data

$\max_k$ = maximum predicted 24-hour concentration within meteorological year k

$\max_{jk}$ = maximum predicted 24-hour concentration within quarter j within meteorological year k

$\hat{p}_i$ = average of highest predicted concentrations from each year modeled with the l years from which meteorological data are used ($\geq$5 years for off-site data, $\geq$1 year for on-site data)

p_{ijk} = modeled daily 24-hour concentration at receptor i, in quarter j and meteorological year k

p_{ik} = modeled daily 24-hour concentration at receptor i, in meteorological year k

r_m = concentration rank of b_{im} corresponding to 98th percentile of all b_{im} in year m, based on number of background concentration measurements per year (n_m). r_m is given by the following table:

n_m	r_m
1-50	1
51-100	2
101-150	3
151-200	4
201-250	5
251-300	6
301-350	7
351-366	8

K.6.4 Equation Set 3: 24-Hour PM$_{2.5}$ design value (Second Tier Analysis)

Formulas

$$\hat{c}_i = \sum_{m=1}^{3} \frac{c_{im \bullet r_m}}{3}$$

$$c_{im} = \forall c_{ijm} \in m$$

$$c_{ijm} = b_{ijm} + \hat{p}_{ij} \text{ , for the eight (8) highest } b_{ijm} \text{ in quarter } j \text{ in monitoring year } m$$

$$\hat{p}_{ij} = \sum_{k=1}^{l} \frac{\max_{jk}(p_{ijk})}{l}$$

Definitions

b_{ijm} = daily 24-hour background concentration at receptor i, during quarter j in monitoring year m

$\hat{c}_i$ = 24-hour PM$_{2.5}$ design value at receptor i

c_{ijm} = The set of all sums of modeled concentrations ($\hat{p}_{ij}$) with background concentrations from quarter j and monitoring year m, using the eight highest background concentrations (b_{ijm}) for the corresponding receptor, quarter, and monitoring year.

$c_{im} = \forall c_{ijm} \in m$ = the set of all c_{imj} corresponding to monitoring year m

$c_{im \bullet r_m}$ = predicted 98th percentile total concentration from the project, nearby sources, and background measurements from year m. Given by the value of c_{im} whose concentration rank in year m is r_m, using background measurements from year m.

i = receptor

j = quarter

k = year of meteorological data

l = length in years of meteorological data record

m = year of background monitoring data

$\max_{jk}$ = maximum predicted 24-hour concentration within quarter j within meteorological year k

p_{ijk} = Predicted daily 24-hour concentration at receptor i, during quarter j, based on data from meteorological year k

$\hat{p}_{ij}$ = Average highest 24-hour modeled concentration (p_{ijk}) using l years of meteorological data

r_m = concentration rank of c_{im} corresponding to 98th percentile of all c_{im} in year m, based on number of background concentration measurements per year (n_m). r_m is given by the following table:

n_m	r_m
1-50	1
51-100	2
101-150	3
151-200	4
201-250	5
251-300	6
301-350	7
351-366	8

K.6.5   Equation Set 5: 24-Hour PM₁₀ design value

Formulas

$$\widetilde{c}_i = \widetilde{b}_i + \widetilde{p}_i$$

$$\widetilde{b}_i = \max_{in}(b_{in})$$

$$b_{in} = \bigcup_{m=1}^{3} b_{im}$$

$$\widetilde{p}_i = p_{il\bullet r_l}$$

$$p_{il} = \bigcup_{k=1}^{l} p_{ik}$$

Definitions

$\widetilde{c}_i$ = 24-hour PM₁₀ design value

$\widetilde{b}_i$ = maximum monitored 24-hour PM₁₀ background concentration at within b_{in}

b_{im} = the set of all monitored 24-hour PM₁₀ background concentrations at receptor i within monitoring year m

b_{in} = the set of all b_{im} within monitoring years n

i = receptor

k = year of meteorological data

l = length in years of meteorological data record.

$\max_{in}$ = the maximum monitored 24-hour background concentration at receptor i within monitoring years n

n = the set of all years of monitoring data, $m = \{1,2,3\}$

$\widetilde{p}_i = p_{il\bullet r_l}$ = modeled 24-hour PM₁₀ concentration with concentration rank of r_l among all concentrations modeled using l years of meteorological data

p_{il} = set of all modeled 24-hour concentrations at receptor i across l years of meteorological data

$r_l = l + 1$ (for example, $r_l = 6$ when using 5 years of meteorological data)

$\bigcup_{a=1}^{z} c_a$ = the set (finite union) of all c_a with integer values of $a = \{1, ..., z\}$